P9-AQN-499

Physics as Metaphor

Physics as Metaphor

by

Roger S. Jones

with drawings by Mike Norman

University of Minnesota Press • Minneapolis

Published by the University of Minnesota Press,
2037 University Avenue Southeast, Minneapolis MN 55414
Printed in the United States of America

Library of Congress Cataloging in Publication Data

Jones, Roger S. (Roger Stanley), 1934-
Physics as metaphor.

Bibliography: p.
Includes index.
1. Physics—Philosophy. I. Title.
QC6.J62 530'.01 81-16496
ISBN 0-8166-1076-2 AACR2

For my mother,
Bella Jones

And in memory of my father,
Bernard Jones

Now I a fourfold vision see,
And a fourfold vision is given to me;
'Tis fourfold in my supreme delight
And threefold in soft Beulah's night
And twofold Always. May God us keep
From Single vision & Newton's sleep!
—William Blake

A world ends when its metaphor has died.
—Archibald MacLeish

Preface

In this book, I offer an idealistic reevaluation of the physical world. I reject the myth of reality as external to the human mind, and I acknowledge consciousness as the source of the cosmos. It is mind that we see reflected in matter. Physical science is a metaphor with which the scientist, like the poet, creates and extends meaning and value in the quest for understanding and purpose. My intention, therefore, is to reformulate some notions about our world by taking serious account of the creative role of human consciousness.

Along the way, I shall draw most of my examples and arguments from physics and mathematics. These are the fields that I have studied, taught, and loved the most, and in terms of which I first began to articulate the issues that have crystalized into this book. I have always tended to visualize and express questions of human meaning and existence in terms of physical and mathematical analogies. Naturally, my education and experience as a physicist have reinforced this way of thinking. But the great power of these analogies has taught me that metaphor is the heart of the apparent dichotomy between reality and consciousness. Thus I use the language and ideas of physics, not as do most scientists, but figuratively, so as to explore the meaning, value, and purpose of human existence.

I have never been able to separate matters of philosophy, belief, aesthetics, and ethics from the study of physics. Within physical and mathematical ideas, it is always the philosophical, the aesthetic, the psychological that I savor and which excite me the most. I believe that many scientists share and relish these tastes with me, but when they exclude all subjective, metaphysical, and ethical matters from their work and communications, I part company with them. I am no longer convinced that this exclusion makes one's work more objective. Nor do I believe that the philosophical approach is too impure and that it will weaken and destroy physics. On the contrary, physics was born in the philosophical quest for meaning and understanding, and rather than doing it any harm, this quest can restore the ancient vitality of physics and regenerate its humanity.

In an effort to embody consciousness and subjectivity within physics, this book is divided into three parts. The first two chapters attempt to debunk the myth of objectivity in science by unearthing the essential subjective core of the process of measurement and by arguing that prediction and generality in science result from a kind of "stacking of the deck." The central chapters examine the "cardinal metaphors" of physics: space, time, matter, and number. Although the scientific and mystical approaches to these metaphors are different, there is no real boundary between them, as examples from astrology, music, language, mathematics, and physics illustrate. The last three chapters explore the motives and ethics behind our physical metaphors, and present some puzzles and suggestions to ponder for the future, all by way of helping us to assume our unaccustomed role as participant-creators of the cosmos.

Many people have contributed to my work on this book in the past six years, and I should like to express my deepest gratitude to them:

To George Boody, Michael Carlston, Joseph Friedes, Robert Hoffman, Richard Hruby, Michael Sullivan, and Paul Tungseth who have read all or major portions of the manuscript and have given me invaluable critical commentary on it.

To my cousin, the late Lee Foster, whose encouragement and professional advice on writing, editing, and publishing were and remain precious to me.

To Richard Clugston for his challenges and questions which have decisively influenced this book, and for his enduring support.

To David White for his unflagging interest, encouragement, generosity, and trust, for his detailed commentary on the manuscript, and for the inspiration of his own writing and teaching.

To Owen Barfield, whose visit with me in 1972 was a vital inspiration, and whose written works are the spiritual ancestors of this book.

To Richard Abel for his faith in me and for his skillful, diligent work on the manuscript.

To Mike Norman, whose sympathetic and creative response to my work produced the witty drawings that grace this book.

To Henry Geiger for his help and advice, and for the synchronistic accord of his wonderful publication, *Manas*.

To the members of my men's support group for their strength and for spurring me on.

To all the friends, students, and acquaintances, too numerous to mention, whose suggestions have greatly improved this book.

Finally, to my wife, Louise, for her professional editorial assistance, and for her advice, patience, trust, and love.

Table of Contents

Physics as Metaphor

Introduction

In the fall of 1967, I finally confronted my uneasiness about science and reality. I had just settled in Minneapolis to teach physics at the University of Minnesota and to work in a large educational project that was developing a new math-science curriculum for the elementary grades. I found myself thinking hard about why and how to interest children in science, and this in turn awakened several philosophical issues that had troubled me over the years. As a practicing physicist, I had always been vaguely embarrassed by a kind of illusory quality in science and had often felt somehow part of a swindle on the human race. It was not a conspiracy but something more like the hoax in *The Emperor's New Clothes*. I had come to suspect, and now felt compelled to acknowledge, that science and the physical world were products of human imagining—that we were not the cool observers of that world, but its passionate creators. We were all poets and the world was our metaphor.

My own idea of metaphor has undergone a transformation. Ordinarily, a metaphor suggests an analogy or likeness between two different things by applying the term for one to the other. For example, in the expression *electrifying news*, the emotional impact of a report is likened to a physical shock. But there is more to the matter than just a transfer of meaning. As Colin Turbayne points out, "the use of metaphor involves both the awareness of duality of sense and the pretense that the two different senses are one."[1] The author and recipient of a metaphor connive, as it were, in agreeing to a tacit *as if*. There is an act of creation at the heart of metaphor which distinguishes it from simpler, more passive comparisons, and explains its essential value in the arts. In my own thinking, I have come to emphasize this active, creative character of metaphor over its more usual passive, comparative quality. I define metaphor as an evocation of the inner connection among things: It is an act of consciousness that borders on the very creation of things, blurring the distinctions between them, even between them and their names. I think of metaphor in the larger context in which all things are related to one another through an underlying unity—some cosmic principle or force, such as Brahman in the Hindu pantheon. Metaphor tantalizingly evokes this hidden synthesis through its dramatic juxtaposition of apparently unlike things. It is as if the poet, in using a metaphor, not only refers to a symbolic or ironic connection between two things, but hints at the very creative act itself which underlies all naming and evoking processes.

For me, all of this is related to the Eastern notion that the cosmos is a seamless unity—the One—which contains within itself, in potential form, the myriad actualizations of the universe, all that which is possible or impossible, all that is or might be. And we are part of some aspect or phase of the One in which only a portion of all the potentialities appears actualized. For us there seems to be existence, structure, difference, space, and time. We experience the manifestation with its variety and multiplicity, but we sense only dimly the unity and often lose sight of it altogether. We forget that we are One and believe instead that we are individuated. Now and then, a seer or poet among us keenly divines the One and tries to urge us toward it once again. In language and poetry, this is beautifully done with metaphor, which reveals to us the unbroken, synthesizing thread that runs through all the varied stitches, colors, and patterns of the cosmic weave.

In discussing physics as metaphor, I suggest that the metaphoric act transcends language and vitalizes creativity in science. It is generally acknowledged that scientists often use creative analogies and comparisons to extend theories and even to make new ones. Modern elementary particles are extensions and elaborations of the Greek *atomos*, and William Harvey found in the centralized sun of Copernicus a model for the heart as the center of the circulation of blood. But I go beyond these relations in form alone and suggest that scientists (and indeed all who possess creative consciousness) conjure like the poet and the shaman, that their theories are metaphors which ultimately are inseparable from physical reality, and that consciousness is so integral to the cosmos that the creative idea and the thing are one and the same.

What else are we to think when the theory of relativity teaches us that space and time are the same as matter and energy, that geometry is gravity? Is this not an equating, an integration, of mind and matter? Is this not an act of poetic, perhaps of divine, creation? And what of the astronomer's black hole, the perfect metaphor for a bottomless well in space from which not even light may escape? Which is the reality and which the metaphor? And what of quarks, the claimed ultimate constituents of matter, locked permanently within the elementary particles they compose, never able to appear in the literal, physical world? Are they not constructs, figments of the mind, symbols for a collection of unobservable properties? How is the quark more real than figurative? And is not the very term *quark* coined from that most metaphoric and creative of works, *Finnegans Wake*? And when physicists, with tongue in cheek, apply terms like *color* and *charm* to the quark, can we believe that they are oblivious to their own creative acts? Indeed, as Sir Arthur Eddington said in 1920, the footprint we have discovered on the shores of the unknown is our own.

For the full elaboration of the idea of science and the physical world as a construct of the mind or a collective representation, I owe a great debt to Owen Barfield and his writings, especially his book *Saving the Appearances—A Study in Idolatry.* It was Barfield who helped me most to fathom the deceptiveness of science by seeing that when metaphors become crystallized and abstract, cut off from their roots in consciousness, and forgotten by their creators, they become idols. For an idolator is not so much one who creates idols, but one who worships them. This failure to recognize the central role of consciousness in reality and thus to treat the physical world as an independent, external, and alien object has been a chronic problem throughout the modern era of scientific discovery, since the Renaissance and the Enlightenment, and has reached a critical stage in the twentieth century with its unconscionable, and largely unconscious, ravaging of the environment.

It is often pointed out that in modern physics, in relativity and quantum mechanics, consciousness has already been brought into the picture. In relativity, the quantitative character of the physical world depends on the frame of reference or state of motion of the observer. Here the language, through no fault of Einstein's, is somewhat suggestive and misleading. For the words *frame* and *state* do not refer to any condition or quality of the mind of the observer, but rather to the system of coordinates by which he or she measures and observes physical phenomena and to the velocity of that system relative to those of other observers. Furthermore, the observer need not be a sentient, thinking being. A collection of automated measuring devices, manipulated and controlled by an electronic computer, would do as well, or perhaps better, as the remarkable outer space observer-rockets of the 1970s already demonstrate.

In quantum mechanics, the situation is more than suggestive. Werner Heisenberg's Uncertainty Principle and its elaboration into a full-blown philosophy by Niels Bohr and others give to the observer an essential role in nature. According to this view (the Copenhagen interpretation of quantum mechanics), physical theory deals only with an observed world (an unobserved world cannot be measured or described and is therefore meaningless to the empirical scientist), and the observer has an uncontrollable and nonremovable effect on what is observed. In other words, the observer and observed form an integrated unit that cannot be broken down into independent components. The precise nature of the connection between mind and matter remains somewhat elusive, however. Is it the very act of observation that causes the interference (if one can speak of causality in this context), or just the mere presence of a sentient being? As with relativity, perhaps only an automated device need be the observer. But ingenious puzzles and paradoxes, dreamed up by philosophers and physicists, seem to displace the threshold of observation ever deeper within the sensory and nervous system of the observer, so that the essential degree of sentience or consciousness remains a somewhat open question. Quantum mechanics, then, may just possibly imply an essential role for consciousness in the scheme of things, but it is basically a passive role and not an active or creative one. Behind all is the assumption that consciousness along with matter is a random, accidental occurrence in the universe without rhyme or reason.

Furthermore, the real issue is whether or not such ideas figure significantly in scientific research. It is, in fact, the rare scientist who is concerned with such matters. The Copenhagen interpretation may be the prevailing philosophy of quantum mechanics today (and it has some serious rivals), but it is hardly a hot topic over lunch at the research lab. Most scientists take a rather pragmatic and condescending view of philosophy, and its niceties have no direct bearing on their day-to-day research, thinking, and discussion. When pressed, most scientists will acknowledge the relevance of philosophy to physical science, but their official pronouncements to the lay public and in their classrooms contain little philosophical leavening to lighten and make more palatable the dense dough of scientific thought. Fifty years after the Copenhagen interpretation forced consciousness on an unwilling scientific community, there is precious little to be found in the research literature of physics to suggest any bridging of the mind-body gap.

In fact, in the last fifty years, the trend in mainstream physical science has been away from consciousness and holism and toward the mechanistic and divisible world of the nineteenth century. Fritjof Capra[2] argues that despite the much touted promises of an ultimate unification in physics, modern elementary particle and quark theory is basically a throwback to the atomistic, thing-oriented notions of premodern physics and is contrary to the holistic, process-oriented currents in modern thought.

Despite the claims of Niels Bohr's influence on physical thought and the arguments of a small number of philosophically minded scientists, references in physics to consciousness are often a matter of lip service. Nor have the profound writings of such philosophers and historians of science as Michael Polanyi, Thomas Kuhn, and Gerald Holton had much effect on thought and research in physics. The child, science, has outgrown and rejected its parent, natural philosophy, which languishes in a slow death.

Thinking about these matters in the fall of 1967, it became clear that the agonizing rift between the arts and the sciences, or between mind and matter, that had plagued me since childhood was false and contrived. My troublesome inability to separate the "two worlds," to concentrate on pure science and to exclude questions of meaning and value from the study of physics, suddenly seemed more of a blessing than a curse—a kind of life preserver, thrown to me as I thrashed wildly about in a sea of illusion.

My view of a person's relation to the physical world can be illustrated by a classic movie cartoon: A frenzied underdog character is discovered in flight, hotly pursued by a menacing bully. Underdog is suddenly frustrated when his path abruptly ends at the top of a precipitous cliff. There is no hope of escape. Bully, gleefully sensing victory, gears up his pursuit of the doomed victim. Underdog is frantic and desperate. Unthinkingly, he begins to clamber up the side of a passing cloud, shaped like a staircase. Bully is totally mystified by Underdog's miraculous deliverance. He stops dead in his tracks and stares at the impossible feat. This provides Underdog with a fatal moment of pause. He looks down, sees to his horror that his feet rest on nothing but vapor, and only then proceeds to fall helplessly to earth. It isn't illusion, but its recognition, that does him in.

In cartoons, there is a delightful ambiguity about reality. Objects often seem like those of the real world: external, inert, subject to the laws of inanimate matter. But occasionally, the characters can interact psychically with these objects, controlling their behavior and participating imaginatively in their existence. They may suspend boulders in air, camouflage themselves by becoming trees, travel instantaneously from place to place, or "pour" themselves through a hole in a fence. We accept all these lapses and contradictions as part of the cartoon world. It isn't real; it's an image, an illusion, created at the whim of an artist.

As long as our own participation in creating the physical world remains unconscious, that world continues to sustain us. But, no sooner do we acknowledge our unaccustomed creative role, than the supporting scaffold evaporates, and we fall. It's as if our choice were between idolatry and damnation. Our mapping of and control over the physical world has so beguiled us that we no longer sense, as did medieval and earlier people, any connection between

In cartoons, there is a delightful ambiguity about reality.

sense, as did medieval and earlier people, any connection between the psychological, spiritual elements of our inner world and the inert, mechanistic elements of the external one. Nor do we see them as different aspects of an underlying unity. But the successes of science neither establish nor guarantee the validity of its central materialistic dogma. Although we may control matter, and treat it objectively, we may still suspect that it is created and sustained by the human mind.

Modern people tend to discount unfamiliar worlds by carefully distinguishing between them and their physical reality. Dream, myth, literature, even modern art are relegated to the realm of imagination and subjectivity, which is assumed to have no objective existence, no consistent physical properties, no logical, eternal, governing laws by which we may predict and control events. We may appreciate other subjective worlds, find them amusing, even derive wisdom from them, but we need not take them seriously or treat them as real and objective. The two realms are separate and distinct—the one exists as part of the mind, the other, altogether apart from mind.

By subjectivity I am not referring to the effects on scientific thought of the individual tastes, preferences, and prejudices of scientists, which change with time, are influenced by peer pressures, and figure prominently in the formation of scientific paradigms. Rather, I mean the basic role that mind and the self play at some unfathomable level in the workings of the universe. Subjectivity in science has both a personal and an impersonal aspect, and fundamentally I mean it to refer to the dependence of the physical world on consciousness. Mind and matter are not separate and distinct, but form an organic whole, in my view. To distinguish a subjective from an objective viewpoint is ultimately illusory.

The crisis in my own thinking was part of a general awakening that swept through Western culture in the late sixties. We had reached a sobering stage in the perennial quest for meaning, where we could no longer look to science for answers. The ancient and respectable search for the nature of space, time, and matter; for the purpose of life and consciousness; for existence beyond death, which had been denounced and banished by science as superstitious and anthropomorphic, has returned from its long exile. It will no longer do to discredit this search as the wishful thinking of confused and unrealistic minds.

Despite the towering intellectual and technological achievements of twentieth-century science, its spell over us has been irreversibly weakened. There are at least two important reasons for this. First, scientist and layman alike have become acutely aware of the limits and shortcomings of scientific knowledge. Second, we realize that our perpetual hunger for spiritual understanding is real and undeniable. It can neither be defined away by subtle logic nor be satisfied by viewing the universe as sterile, mechanistic, and accidental.

I believe that the human search for meaning and value is of paramount importance, and that physics can shed light on that search, provided first that it stops masquerading as an objective body of knowledge and reveals its subjective nature. By probing the human and imaginative aspect of physics in this book, and breaking down its false subject-object barriers, we shall find new life in science. We shall see that the celebrated ability to quantify the world is no guarantee of objectivity and that measurement itself is a value judgment created by the human mind. The scientized concepts of space, time, matter and number will be explored as metaphors, expressing the human need and ability to create meaning and value. These metaphors have an intuitive, mythic, life-giving character which completes and enhances their quantitative meaning and which is motivated by basic human fears and yearnings. It is the fear of death and chaos that gives the metaphors of science their modern lifeless and alien character. But we can restore life to our metaphors and meaning to our lives by renouncing scientism and accepting responsibility for our own creative part in the cosmos. Then, perhaps, we shall be able to frame the metaphors of the future through a conscious, moral act of human imagination.

This book is an exploration of the subjective side of the metaphors of physics—their metaphysics, psychology, and ethics—with the dual purpose of illuminating the human quest for meaning and reevaluating the role of science. To lay the groundwork for this, it is first necessary to discredit the dogmatic belief we hold in the objectivity of science. It is useless to seek the seeds of space and time in the human mind without first loosening the tenacious hold that three centuries of scientific indoctrination have had upon us. We are too strongly convinced of the objectivity of space and time to contemplate them seriously as our own poetic creations. The first two chapters undertake this task of iconoclasm. They argue that objective measurement is subjective by its very nature and that science's ability to predict and generalize stems from the stacking of the scientific deck of knowledge. This case for the illusion of scientific objectivity attempts to make way for the scientific mythos to which the main body of the book is devoted and may be skipped over by those already aware of a certain idolatry in science or who are keen to sample metaphoric and visionary physics.

Chapter 1.

The Lengths We Go To

In Alexander Korda's classic film, *The Thief of Bagdad*, there appears a wondrous magical ruby, called the all-seeing-eye. It has the visionary power to conjure up and display any view requested of it. Peering into its crystalline, cardinal depths, one could see at will anyone, anywhere, anytime. This charismatic film entered deeply into my childhood fantasies. I identified with the hero of the story, played by Sabu, a famous boy star of my generation. Often I would replay in my mind Sabu's daring theft of the ruby from the forehead of an enormous, foreboding statue of the Buddha, guarded over by a huge hideous spider. In disbelief and awe, I would recall how Sabu, in a fit of anger and frustration, brought on by the all-seeing-eye's revelations, had flung the ruby down a steep rocky escarpment in the mountains, and how its shattered fragments had sent up great billows of smoke in testimony to its uncanny powers, now lost forever.

After seeing the film, I found a dimestore ring with a red rhinestone in it. This became my all-seeing-eye. I carried it with me secretly, drawing sparingly of its magic as I lived my own Arabian adventures.

Even today, as I think back on that ruby, I feel a clutch of inexplicable enchantment and attraction.

My recollections of that bewitching gem epitomize for me the scientific folly of the quantitative. All my associations with the film—the magical visions, the exotic Arabian culture, the romantic adventures, the imperious Buddha monument, the terrifying spider—add vibrant color and meaning to the bare memory of the ruby, and make it dazzle all the more within my mind. The thought that such a ruby could possibly be described quantitatively, in the cold clinical terms of some chemist or geologist interested in its mineral content or cleavage planes, is for me nothing short of blasphemy. Pure quantification cowers and collapses before such an apparition. Numbers fail.

The heart of our modern idolatry is quantification—the world reduced to quantities and the relationships between them, the belief that the quantitative description of things is paramount and even complete in itself. Physical science persuades us to disregard, devalue, and even deny what we cannot measure, to act as though such things as love, life, optimism, wonder, and beauty do not matter much in an objective description of reality. I do not dispute that the wealth of knowledge science gives us adds to our wonder and experience. I may even appreciate rubies all the more by knowing something of their chemistry and geology. But I rebel at the dominance of quantitative description in our scientized lives which sustains our idol of objectivity and keeps us from an intimate participation in our world.

In this chapter we shall explore in some detail the process of measuring the length of a table in order to expose the fundamental and irreducible subjective nature of measurement itself. The reason for all this fuss is that science's claim to objectivity rests ultimately on its ability to make exact and reproducible measurements. That the sun is 93 million miles from the earth and that the human body is normally at a temperature of 37°C seem to us matters of fact, unaffected by the subjective world of dreams and myth. There is something about the measured size of an object that seems to guarantee its objective status. If a table is three meters long, what could that possibly have to do with the human mind or how could that be affected even by the annihilation of the entire human race?

Our idolatrous notion of an objective world, independent of our minds, is sustained by the qualitative, subjective feel of our senses, corroborated by objective measure. The raw data of my senses is largely *interpreted* by me in terms of a construct of reality based on the measurement of physical properties. I do not see a table. What I perceive is an oddly shaped brown area in my visual field. I do not see its height, breadth, and width. I infer these by fitting my touch and visual sensations into a predetermined construct for a table which has certain spatial properties with measurable dimensions. I cannot even feel the thickness of the tabletop or its solidity. I feel only a sensation in my fingers (called pressure) when I try to close them, or an abrupt resistance to my fist swung down before me. These constructs are bolstered by the knowledge that I can measure, and therefore verify, the table's height, weight, thickness, and so on.

Of course, this interpretation is almost instantaneous and unconscious. It requires great effort and concentration to distinguish that which I actually see and feel from the table I tell myself is before me. Years of conditioning have established and reinforced the physical, spatial interpretations we make—a long process of indoctrination that is extremely difficult to recognize as such and which plays quite a useful role. It makes possible the hand-eye coordination and integration of sense data that enable us to live an efficient, ordered life. Without that, what would our world be like? A chaos, unintelligible, unnegotiable; a schizophrenic nightmare, the experience of an infant mind, or perhaps even of death.

It is with the aim of establishing and illustrating the subjective, metaphorical element in our physical reality that we shall, in the main portion of this book, turn to an exploration of the four basic scientific constructs, space, time, matter, and number, which I call the cardinal metaphors. We shall probe the subjective, mythic, imaginative character of these metaphors and see that their quantitative and qualitative sides are inseparable, and that there is an essential human element even in the coldest, most objective and external of scientific concepts. At the outset, in this chapter, we shall find that the quantitative description of things not only is incomplete but is, at its heart, ultimately subjective. The citadel of objectivity conceals within its foreboding material walls a vast treasure of human imagination and creativity. But its deepest secret is that its hidden foundation is embedded in the human mind. So before moving on to the marvels of the upper halls and turrets, we shall poke around a little in the basement.

Measure for Measure

It is an amazing fact about physics that none of its concepts are ever really defined. What we are given instead of a definition is a prescription for measurement. To build a rocket and send it to the moon, you need only measure space, not define it. The measurement of space is the only specification of it needed for scientific purposes, and this is called an operational definition. The remarkable twentieth-century achievements of the physical sciences—television, computers, outer space exploration, and atomic energy, to name only a few—clearly attest to the utility of the notion of an operational definition. Yet it is no definition, but a prescription, a recipe. Telling me how to measure space may make it possible to navigate in it, but it does not tell me what space is.

Operational definitions are in fact passé in modern-day philosophies of physics. Defining physical quantities in terms of a prescription for measuring them was a central idea in the scientific philosophies of logical empiricism and operationalism. Such approaches are largely viewed today as naive, simplistic, and incapable of dealing with the subtleties of late twentieth-century physics. But the arguments among opposing schools of philosophy and even the current trends in science philosophy are not the real issue for us, simply because they are not an issue for most practicing physicists today. In this book, I am primarily interested in how scientists think about science and the physical world, the images and models they use among themselves and utlimately transmit as absolute truth to lay persons, however inadvertently.

Philosophers of science may dispute the meanings and epistemological status of concepts and definitions in physics. They may even leaven science philosophy with historical and humanistic issues. But such considerations are not the currency of exchange in modern physical research, from which popular ideas in science are finally derived. The language of research and its popular counterpart is still replete with the conceptual heritage of Galileo, Descartes, and Newton, with its fundamental assumption of a separate, quantifiable, objective world. Operational definitions may not be in vogue in the philosophy of science, but when a physicist today wants to convince someone of the objective truth of some observation, he or she will use measurement as the trump card.

Despite the feats of modern science, we have no clearer understanding of the nature of space than did our Greek ancestors. By understanding I do not mean any mental construct we impose upon space, such as geometry or perspective. I mean the essence and experience of space itself—the basic mystery in which we all participate, which permeates our every act and thought, and whose ubiquitous presence we accept unconsciously as synonymous with our very existence in the world. Nor does the success of science demonstrate the irrelevance of such metaphysical matters as the nature of space and time. It may be a matter of indifference to science, but not to a human, whose very life and meaning is delimited by the confines of space and time.

But to discover subjectivity in science, we need not invoke metaphysics, but only ask how length is dealt with in physics. You may well wonder why length and why physics. Length, time, and matter form a kind of basic trinity in the exact sciences. All other physical quantities may be defined in terms of these three. For example, the speed of a car is the distance the car travels divided by the elapsed time. Not only are such obvious things as speed and acceleration defined in terms of space, time, and matter, but also things like force, energy, temperature, electric charge, and voltage. The question of meaning in the exact sciences boils down to how space, time, and matter are defined. And they, in turn, are defined operationally in terms of their measurement.

We shall consider only space. Even within the basic trinity, space is a kind of prototype for the other two. Measurements of time and matter depend in practice on the measurement of space. Clocks and balances employ length scales, and even digital clocks depend on the precise size and geometrical arrangement of their components to produce calibrated time signals. It is conventional in everyday physics to use space, time, and matter as the basis of all other measurement, but space alone is sufficient.

As to why we single out physics, it is because physics is basic to the other sciences and is the quantitative science par excellence. Chemistry and geology are quantitative to a degree, but, unlike them, physics deals only with the quantitative aspects of things. Physical laws are expressed in terms of mathematical statements, equations, in which measured quantities appear symbolically as variables. For any concept to appear in a physical law, it must first be expressed quantitatively. Thus we must have a way of assigning a number to the physical concept, and this in effect is what measurement in physics is all about. The operational definition of length is really a prescription for assigning a measure or number to space in a precise and reproducible way.

It might be argued that mathematics is more quantitative than physics. This is not necessarily so. Mathematics treats some topics rather qualitatively, for example, plane geometry and topology. But even though mathematics is usually highly quantitative, its concepts are pure abstract creations of the mind and bear no necessary connection with the physical world. By contrast, physics begins and ends with the real world. In the mainstream view of physics (which I am calling into question), the ultimate value of a physical theory is its ability to describe, predict, and control the phenomena and events of the physical world. The great success and fame of physics rests primarily on its claim that the most useful knowledge about the physical world is obtained by measuring or quantifying that world. Herein lies the objectivity of physics and the basis of our objective view of the world. The final retreat, the ultimate stronghold of objectivity, is the fortress of measurement.

A Fable of Length

I wish to measure the length of a table. I choose to adopt the definition of length used in physics, for, as we have seen, it is an operational definition, telling me *how* to measure length. I take as my standard unit of length the meter (equal to 39.37 inches). To measure length, I must follow three basic steps: 1. Define the object to be measured. 2. Compare the object with the measuring device. 3. Count units of length.

As we shall see, the most essential and irreducible subjectivity is present where least expected, in step 3, in the process of counting itself, which lies at the very heart of measurement and quantification. But we shall begin at the beginning, and take at least a brief look at the problems connected with the first two steps.

STEP 1: DEFINING THE OBJECT

First I must locate and identify two unique points that will form the limits of the measurement. My table is rectangular, and so the two ends of the longer dimension are the points in question, and the distance between these points is defined as the length of the table. Now, if I am measuring for a tablecloth, I can afford to be pretty sloppy about all of this. A tablecloth has enough overhang that 10 or 20 centimeters either way will not matter much. If, instead, I want a glass plate made to fit neatly on the tabletop, I should be more careful. An error of a couple of millimeters is as much as I can tolerate for a good fit. But to understand the definition of length, I should first ask for the most precise possible measurement of the length of the table. This is really asking for trouble.

The problem is that in trying to make an ultimate measurement of the table, we need to think in terms of a microscopic, or even atomic, determination of the limiting points. We now lose the everyday concept of a table as a solid object with well-defined boundaries. Instead, we must use the modern picture of matter as molecules in motion, and we must try to imagine how to define the edges of the table in terms of the rapidly speeding molecules that make it up. To make matters worse, the atom of modern physics is, strictly speaking, not even visible in principle. The atom is not the hard little ball of popular illustrations and high school science texts. It is something of a mathematical abstraction whose description in terms of the spacetime of our everyday world requires some compromise. Today's physicists use such terms as electron clouds and probability densities in describing atoms. These terms have fairly precise mathematical definitions. But their use to represent atoms visually as fuzzy-looking balls, dumbbells, and ellipsoids is usually misleading.

The Heisenberg Uncertainty Principle—the fundamental law on which the modern theory of the atom is based—tells us that we cannot make an exact determination of the position of an atom without disturbing the atom so violently that its whereabouts an instant later will be completely unknown. If I try right now to get an exact fix on an atom in the corner of my table, I shall have no idea, even a fraction of a second later, whether that atom is still in the corner of my table or somewhere in England. To guarantee some reasonable knowledge of the future position of an atom, I must not insist on measuring its present position too precisely. There is always a tradeoff in accuracy between present and future knowledge, or in uncertainty between position and velocity, according to Heisenberg's Principle. What the fuzzy-ball representation of the atom shows is not an actual picture of an atom, but a sort of time-average of a large number of imprecise position measurements. The idea of a snapshot or instantaneous picture cannot be applied to an atom, not even in principle. An atom cannot be represented or conceived in the spacetime of the familiar world.

Heisenberg's Uncertainty Principle is not a statement about the limitations of measuring devices, but rather about the limits to human knowledge. The difficulty of determining the boundaries of an object through the position of its constituent atoms, or, even worse, of determining the boundaries of the atom itself, should we be so foolish as to aim for subatomic accuracy, cannot be eliminated through better instrumentation. The inaccuracies we must settle for at these levels cannot be avoided. They are a built-in feature of the modern theory of matter. Nor can we overcome these problems by making some arbitrary decisions about where the boundaries of an object or an atom are. This is because we cannot define the space an object occupies without reference to that object, and because we should then be making human judgments, however well rationalized, which would incorporate a nonremovable subjective element in the definition of length.

It might be argued that such a level of accuracy is meaningless anyway since a length measurement at the atomic level cannot be performed in practice. But we are discussing the principle of the definition of length. Furthermore, even if we do give up the atomic level for the definition, down to what level does the table have a meaningful length? We see that length is not well defined, even in principle, but has a built-in unspecifiable uncertainty. For the moment, we are safe in making measurements for tablecloths and glass desktops. For the moment.

STEP 2: COMPARING TABLE AND RULER

Before I can perform the essential third step of counting the number of meters that fit along the edge of the table, I must compare the defined interval (the length of the table between its two ends) with the meter stick. This will allow me to decide which points on the meter stick determine the starting and stopping points for the count. To locate these two points, two kinds of decisions have to be made: the first about the coincidence or relative proximity of two points, and the second about the relative position of three points.

We begin by lining up one end of the meter stick with the left end of the table. (This involves a decision about the coincidence of two points). We assume that the meter stick or tape measure is longer than the table,[1] so that all I need do is decide between which two marks on the meter stick the right end of the table lies. (This involves a decision about the relative position of three points.)

In practice, deciding whether two points are coincident boils down to making judgments about the distance between two points. Point A on the table and point B on the meter stick cannot literally be coincident, for two objects cannot occupy the same place at the same time. I should call the points coincident when the distance between them appears to be negligible in comparison with any meaningful distances involved in the measurement, for example, in comparison with the smallest unit of distance on the meter stick. This presents some difficulties. First, the judgment of sufficient closeness or negligible distance, however obvious it may seem, is still a matter of human judgment and thus subjective. Nor can we avoid this problem by making an "objective" measurement of the distance between the points in question with some more precise measuring device. For then we are simply starting the length-measuring process over again at Step 1 and will sooner or later have to return to the problem of judging the coincidence of two points at some magnified level.

Second, judging closeness means estimating or measuring distance. But we are in the process of giving an objective definition of length by means of its measurement. (Recall that this operational definition of length is the only definition of length we have in physics.) We find that we must be able to estimate length in order to measure or define it. A prior understanding of length is needed for the definition. Such a circular definition is no definition at all.

I must emphasize, at the risk of repeating myself, that although this logical problem seems to have no bearing on the practical application of the concept of length (for example, we can measure space well enough to send people to the moon and back), it is of paramount importance in debunking the notion that an objective definition of a measurable length guarantees the independent existence of objects.

The other kind of decision we make in comparing the ruler and table is about the relative order of three points. We must decide between which two marks on the meter stick the right end of the table lies.[2] We have three points, A, B, and C, and we wish to establish that B, the end point of the table, lies between A and C, two particular marks on the meter stick. We note that observing the order of three points does not involve the measurement or estimation of distance. The notion of order belongs to the nonmetric form of geometry called topology, or colloquially, rubber-sheet geometry, which deals with connectedness, continuity, insideness, outsideness, order, and similar concepts independent of the size of things. If we can reduce the measurement of length to purely topological operations, we shall have eliminated the circularity of the definition. But we shall see that this is not possible.

To observe the order of three points, we must be able to identify and label the points. Indeed, we have assumed this in calling B the right end point of the table and A and C the two particular marks on the meter stick that flank B. But who is to identify and label the points? Shall I, or any other subjective being? To make the definition of length objective, we must be able to identify the points without recourse to the powers of recognition of the human mind. Could a T.V. camera and computer in combination,

Could a T.V.-computer combination be programmed to identify any object?

for instance, be programmed to identify any point or object from another? There are pattern-recognition programs today, but they often involve subject matter of limited scope, like the 26 English letters, 10 Arabic numerals, and a few punctuation marks. It might be argued that given enough memory space, a computer could be taught to recognize any object that a human being can recognize. If a person can recognize a battleship or an elephant, why can't a computer? Why not even a computer capable of distinguishing one elephant or one mark on a ruler from another? Certain gestalt psychologists and philosophers of science would disagree. Michael Polanyi, for example, would argue that the whole cannot be analyzed into its parts and that tacit, inexpressible knowledge is always involved in human recognition, which can never be programmed into a computer.

But we cannot settle this ongoing debate. Instead, we can consider how we should provide a computer with the necessary information to enable it to recognize or distinguish two elephants, or two marks on a ruler, or two ideal mathematical points, the latter being the simplest objects conceivable. We should do it, of course, by supplying the most complete *quantitative* description we are capable of making. In the case of the elephant, we should give its size, weight, height, shape, trunk length, number of appendages, some measure of its color, skin texture, chemical composition, the audiofrequency spectrum of its bellow, and so on. But all these quantities necessarily depend on a prior definition of length, which we are in the midst of defining. Circularity rears its ugy head once again. (Or was that an elephant?)

We could try to avoid all the problems of describing an elephant or any complex object by programming the computer to recognize the simplest object of all, an ideal mathematical point. A mathematical point has no properties whatsoever. Only one property can distinguish one point from another—its position in space. But, of course, to specify a position in space, we must assume the measurement of length. We are stuck again. There is no purely objective description of anything that does not depend on a prior perception of length. Even if a program *can* be written enabling a computer to recognize an elephant (which is no easy matter and may even be impossible), that program must assume a prior concept of length. Such a program-cum-computer could not legitimately be used in determining the order of three points. For then we'd be assuming length while defining it. So step 2, comparing table and ruler, requires a prior definition of length and/or subjective judgment.

STEP 3: COUNTING

The matter of assigning a number in measuring length might seem even more trivial and obvious than the two earlier steps. Having found the limit points of the table and having compared them with the ruler, what could be simpler than counting the number of meters that fit along the edge of the table? Even the fractional part of the measurement reduces to simple counting. Deciding whether a table is 3.34 or 3.35 meters long is a matter of counting 34 or 35 centimeters after we have already counted 3 full meters along the edge of the table. Our immediate question is how we know that we have 3 meters and 34 centimeters, or 3 of anything and not 2 or 4 or, for that matter, not 6,000.

Make no mistake about it: counting is the heart of measurement. The assignment of number is what measuring is all about. The preparations of steps 1 and 2 make sense only in the light of the basic assumption that there is a unique correspondence between numbers and things. A table is said to have a length because a basic correlation is assumed to exist between the real numbers (the integers, the fractions, and the irrationals) and the continuum of points on a straight line. We make use of this correlation in practice through counting. And counting in turn depends on the concept of number.

Anyone who has ever thought much about numbers knows how difficult it is to define them. Much of the long history of mathematics revolves around the Herculean struggle to define and generalize the concept of number, as we progressed from natural integers to fractions and irrational numbers, then to negative numbers and finally to imaginary and complex numbers. Only in the late nineteenth century, with the invention of set theory, did it become possible to construct a definition of numbers, both logically adequate and meaningful in practice.

Numbers are defined in modern mathematics in terms of the

properties of sets. A set is any collection of objects, real or imaginary. It can be a group of apples or elephants, or it can consist of an apple, an elephant, a sailboat, and a ghost. Any collection of anything, even of nothing, is a set. It may seem that such a simple and vague idea cannot lead to anything useful, but it is the very elementary and general character of the set notion that gives it its great power of mathematical abstraction and universality. By studying collections of objects in the abstract, we can learn something about the arithmetical properties of whole numbers. For example, by combining several sets together, we can learn the laws of addition, as children did when set theory was introduced into the American elementary school curriculum in the 1950s and 1960s.

To get at the idea of numbers and counting, the mathematician introduces the notion of the *cardinality* of a set. The objects that make up a set are called its members or elements. Two sets are said to have the same cardinality if each element of one set corresponds to a single, unique element of the second set, and if in making this correspondence, we exhaust all the elements of both sets. This is called a one-to-one correspondence. One simply compares two sets, member by member, to see whether they have the same cardinality. A unique one-to-one correspondence means that every element of each set is used once and only once in making the comparison. It should be intuitively clear that when two sets have the same cardinality, they have the same number of elements. But the one-to-one correspondence enables us to compare the cardinality of two sets without actually counting their members, and it is therefore more fundamental than counting in allowing us to understand and define the concept of size and number.

A number is defined as that property which all sets of the same cardinality have in common. The abstract notion of *threeness* is that property which a certain set, A, consisting of a star, a dream, and a monkey, has in common with all other sets whose elements can be placed in a unique one-to-one correspondence with the set A. We can glimpse here the great power of the set approach. It provides a very general definition of number, allows us to distinguish between 3 and 4 without ever having to count anything, and offers children a simple concrete way of learning their multiplication tables.

We are now able to place the numbers in an order or sequence. A set having 8 elements, for example, can be seen to be larger than one having 7 elements because, in making a one-to-one correspondence between the two sets, the first will have one element left over after all the elements in the second have been used up. Note that "7" and "8" in the preceding sentence are used only as names or labels for the cardinality of the two sets. They are arbitrary symbols. 8 is found to be larger than 7 because of the one-to-one correspondence and not because of any preconceptions we have as to the innate size of eightness and sevenness. Both the magnitude, or cardinal property, and the sequence, or ordinal property, of numbers follow from the powerful and fruitful notion of one-to-one correspondence, as do all the laws of arithmetic.

We are ready now to return to the matter of assigning a number to the length of a table. I want to determine how many meters fit along the edge of the table. First I mark the meter segments off along the edge. Then I label each segment with the letters, A, B, C, D, and so on. I am using a one-to-one correspondence even in this labeling process, for the set of labeling letters and the set of meter segments that fit along the table have the same cardinality. To find out whether the set of meter segments, or the more convenient set of labeling letters, has a cardinality of 3 or 4 or 7,108, I need only make another one-to-one correspondence with some standard sets whose cardinality is known. Just for fun, and to make the argument painfully obvious, I shall assume that my standard sets have elephants for their members. The first set, called 1, consists of the elephant named Arnold. The second set, called 2, consists of the elephants named Dorothea and Howard. And so forth. For convenience, my labeling letters are printed on a set of cards which I attach to the meter segments as I lay them off along the table. The only cards I use, as it turns out, are labeled A, B, and C. I remove these cards from the table and, for purposes of comparison, carry them over to my standard sets, taking great pains not to drop a card or pick up an extra one on my way to the great cardinal elephant barn. I try to make a one-to-one correspondence between my cards and Arnold and discover that I have cards left over. I conclude that the cardinality of the set of letters, and therefore of the set of meter segments, is greater than 1. By comparing with Dorothea and Howard, I discover that the table is more than 2 meters long. On the next trial, however, I find that I can place one and only one card on each elephant (the elephants are well trained and each patiently holds a card with its trunk) and that I have no cards or elephants left over at the end. Esmeralda corresponds uniquely to C, Jonathan to A, and Florence to B. I have determined that the table is 3 meters long.

This whole procedure may seem a bit ludicrous, but a little thought will convince you that it, or something like it, is essential for someone who doesn't carry around in his or her head the convenience of a memorized sequence of numbers with agreed-upon conventional meanings. It is also essential for us in trying to get at the heart of the measuring process. Having come this far, it would be unfortunate if I had to base my statement that the table is 3 meters long on my subjective ability to distinguish 3 from 4 without any appeal to some objective standard. We might have used more convenient and portable standard sets. But elephants have the advantage of a certain weighty authority. Once the logic and objectivity of the procedure is established, we may return to the convenience of our own memorized numbers.

We are able to objectify and operationalize counting through the use of the fundamental one-to-one correspondence as applied to sets. Our ability to measure ultimately depends on it. But making a one-to-one correspondence in practice depends, in turn, on the ability to distinguish and identify the separate elements of a discrete set. The notion of a discrete set is very basic. It consists of separate, distinct, nonoverlapping elements. The cardinality of a discrete finite set may always be described by a whole number. Finding the cardinality of a discrete set is thus the prototype for counting and measuring. A bowl of jelly beans is an example of a discrete set, whereas a bowl of pudding is a model for a continuous set, one whose elements are not separate and distinct and cannot be counted with whole numbers.

The discrete set of interest to us consists of Esmeralda, Jonathan, and Florence. Our problem is how to distinguish these elephantine elements from one another. But this is precisely the problem we encountered in step 2, when we tried to determine the order of three points. There we tried to distinguish three points, or three elephants, or three objects of any kind. We found that it was impossible without relying on our subjective powers of object recognition and/or without using an objective description which relied on a prior definition of length. Once more, we need length and/or the human mind to define, i.e., to measure, length.

There is an old Indian fable about six blind mendicants who, in their travels, happen upon an elephant in the forest. None of them has ever encountered an elephant before, and the first one, bumping into the trunk,cries out, "Get back; it's a snake." The second, coming from the rear and catching hold of the tail, laughs and says, "No, it's only a rope." The third man collides with the elephant's leg, concludes it's the trunk of a tree, and wonders what the others are talking about. The fourth, clutching the elephant's ear, is convinced he's run into the great fleshy leaf of some tropical shrub. And so forth. Even more to the point, we might enlarge the fable and have the six mendicants encounter not one elephant, but a herd of them, milling about in the jungle. Assuming they aren't trampled, imagine the problems these blind men have now. Not only can't they recognize the nature of the beasts, but they can't even distinguish one from another, or tell where, in the midst of the rushing herd, one elephant ends and the next begins. We find ourselves in a similar predicament. The very ability to recognize objects and to distinguish them from one another requires the full faculties of the human mind and sense organs.

Thus, counting the meters in the length of a table is based on the idea of a one-to-one correspondence between the elements of discrete sets. This in turn depends on distinguishing and separating the elements of a discrete set, which, in practice, requires the powers of human recognition and/or an objective description involving the measurement of length. We cannot define length without already knowing what it is.

The Visit

While writing the preceding section on counting and cardinality, I had an unusual experience. Looking up momentarily from my typewriter, I glanced out of the window and saw a beautiful red bird alight in my backyard. You guessed it—a cardinal. It was late in the year for cardinals in Minneapolis and preoccupied as I was with my writing, I did not immediately notice the connection between the beautiful bird and the central idea in my paragraph. But as my mind returned to my subject, I suddenly realized that both it and the bird had the very same name—cardinal.

Such an occurrence might seem to be a mere chance coincidence. But to me, it was a remarkable synchronistic event, to use the term of Carl Jung. *Synchronicity*, according to Jung, describes the coincidence of two events that have no causal connection between them, but which have a common symbolic or metaphorical content. Jung discusses many amazing examples of synchronicity in his own and other people's experience. Sometimes they occur in dreams and sometimes in the waking state. They need not necessarily even coincide in time. I have experienced a few myself, and I always feel very grateful for them. For they have the quality of a rare and precious gift. This recent one was a kind of gentle, whimsical hint that I am on the right track—a little encouragement from the powers that be in the form of a visual pun.

And such a pun, or metaphor, is particularly apt. It reminded me that even the most abstract notion of size and quantity, cardinality itself, has another meaning. What else is it? A magnificent bird, a prince of the Roman Catholic Church, a deep rich red, an astrological characterization of the pioneering spirit? Do these all have something in common? Or is the total concept of cardinality, like that of elephant, more than the sum of its component parts? Used as an adjective, *cardinal* implies of prime importance, chief, principal, or fundamental. We speak of cardinal significance, of the cardinal points of the compass, of the cardinal virtues and sins. Our English word is derived from the Latin *cardo*, a hinge, thus implying something on which other things hinge, something fundamental. According to the Oxford English Dictionary, the usage of such phrases as "the cardinal winds" and "the cardinal points of the compass" has led to a certain association of the adjective with the number four. (The cardinal virtues and sins were later increased to seven.) There are also the four cardinal signs of the Zodiac: Aries, Cancer, Libra, and Capricorn. And the number four, and the square associated with it, has, according to Jung, the archetypal meaning of a completion, a totality, a return to or echo of the primal unity.

Now it is fundamentally true of an archetypal symbol that it cannot be completely defined and described. Intellectually, verbally, it can never be encompassed. It is best to approach such potent metaphors in a contemplative and open frame of mind, rather than analytically. So one does best simply to stand before this whole matter in wonder and awe, trying to fathom, but not too consciously, some inner connection among the ideas or some deep synthesis of amount, fundamentality, and wholeness.

Chapter 2.

The Stacked Deck

Let's imagine a conversation taking place about a hundred years ago between Len, a layperson, and Ed, an engineer, who is designing a railroad system for the United States. Len has not heard about this before and is intrigued. Ed talks about his plans to lay track between all the major cities in the country so that you can travel easily from one to another. He becomes increasingly enthusiastic and animated as he describes the wonders of his new system.

Ed: You just get on the train in Atlanta, and the next thing you know, you're in Washington, D.C.

(Len looks a bit skeptical and opens his mouth to say something, but is cut off by Ed's exuberance.)

Ed: Well, naturally, you don't get there immediately; that trip would take about a day. But you can forget that you're traveling. The train's got everything you need. You eat and sleep in real style. You just relax, enjoy yourself, and rest assured that you'll arrive in Washington right on time.

Len: On time? How do you mean?

Ed: Ha ha. That's what's so great. You see we're going to publish a time schedule. No matter where you leave from or where you go, it will tell your exact departure and arrival time. You'll be able to read this timetable, as we call it, and tell how long it takes to get from New

York to Philadelphia, or anywhere else. Those trains will be so perfectly timed, it'll be like making an absolutely safe prediction. You'll be able to account in advance for every moment of your journey; even put money on it if you want to.

Len (impressed but thoughtful): Sounds like a fine system of transportation you're designing. It would be great to be able to predict my travel time like that. I get around a lot. But let me ask you this. I have an elderly aunt I've been wanting to visit for some time. She lives on a farm outside of Bryce, Illinois, twenty miles or so from Rantoul. Would I be able to tell from your timetable when I'd be arriving at her house?

Ed (taken aback; looks oddly at Len): Now hold on. That's a little unreasonable. We couldn't predict anything like that, let alone get you to your aunt's place by train. Err, does she live in a town? We can't plan to stop near places like Rantoul, or that—what is it?

Len: Bryce.

Ed: Yea, Bryce. You'd have to get off in Chicago and travel by coach or horse. I have no idea how long it would take. You can't expect a big expensive public railroad to stop in every little hick town. No offense to your aunt.

Len (with dawning awareness): I see. You can provide me with transportation and give me a perfect travel schedule, so long as I go to places you decide are important. I'm not so sure about your railroad after all. What about all the people who live in rural America, far from your big-city railroad stations?

Enough of this hypothetical conversation. The railroad designer would go on to point out that even though most Americans don't live in big cities now, they will soon enough. But that's a whole story in itself.

Physical science, with its remarkable self-consistency and predictive capacity, constitutes a closed, self-defined system like the railroad in our little anecdote. I now want to examine this second essential aspect of science's renowned objectivity.

In the previous chapter, we explored the measurement or operational definition of length in physics and found it to be essentially nonobjective. It always requires acts of human judgment and discrimination which cannot be eliminated without making the definition circular. The entire quantitative description of the external physical world was seen to rest ultimately on subjective human values. Now if the case against the objectivity

You can't expect a big expensive public railroad
to stop in every little hick town.

of measurement is accepted as logically valid the principal argument against it is that it is irrelevant: Even if there is a rotten core in the apple of scientific objectivity, it can only be of importance to philosophers, since physics clearly works. It has unleashed the energy of the atom and placed people on the moon. It enables us to predict and control the phenomena of nature with remarkable success. So what does it matter if there is some logical fallacy in the definition of objective length?

This compelling and pragmatic claim is based on two powerful characteristics of exact science: its predictive ability and its universal generality. I shall now argue that these characteristics result from a kind of stacking of the scientific deck. Games played with this deck must follow a predetermined pattern, just as the scientific Pullman can follow a predictable schedule only where track is already laid.

As an example of a theory with great predictive power and universal validity, I have selected the granddaddy of them all, Newton's Law of Universal Gravitation. One thing that modern and primitive cultures seem to have in common is the special status they assign to people who can divine and anticipate the course of natural events. When such powers include prediction and control, those who possess them are thought to have an almost mystical quality. The modest twentieth-century scientist would disclaim any such special status and attempt instead to show you "how it's done." But the years needed to master the scientific method are a severe hindrance to lay understanding. Even today's uninitiated tend to place scientists, if only unconsciously, in a kind of priestly class.

Imagine, then, the dramatic impact on the seventeenth-century mind of Isaac Newton's masterwork on the solar system. That marvel of mechanical analysis began with a few general and fundamental principles, the laws of gravity and of motion, and derived from them, for the first time, a mathematical description of the pattern of motion of the planets, which had been stated in its elegant modern form only 75 years earlier by the great astronomer Johannes Kepler. The elusive pattern of the planets in their periodic movement against the background of the fixed stars had defied the human attempts to describe it accurately for thousands of years, ever since the time of the ancient shepherds who peered in awe at the night sky, and first descried its puzzling regularities. Kepler had succeeded in reducing thousands of years of observation and complex description to a few simple, precise statements. One of these statements is that the planets move along the path of an ellipse with the sun at its focus. Newton showed that this hard-won, but purely geometrical, description follows inexorably from a set of physical principles so general that they describe not only the regularities of the heavens but also the seemingly chaotic behavior of phenomena on earth, thus destroying even the Aristotelian dichotomy between the heavenly and earthly realms.

We are still reeling today from the impact of Newton's great synthesis. His remains the prevalent model for scientific theory. And although Newtonian physics has been eclipsed by relativity and quantum theory, Newton's use of a few general principles from which to derive a precise description of large classes of phenomena continues to dominate not only physical science but even biological and social science. Newton's method, even more than his deciphering of the planetary order, is the intellectual legacy of the Enlightenment. And at the core of this method is its mathematical predictive capacity.

Predicting the Predictable

Every now and then, we are informed that at some time and date in the near future and at certain locations on the earth, there will be visible that most spectacular of astonomical events, a total eclipse of the sun. Such information is usually accepted without question. This is no mere probabilistic claim that the stock market will rise or fall or that so-and-so will be the next president of the United States. This is a sure thing. No sane bookmaker would bet one cent against the eclipse, and thousands of people, including many scientists, travel from all parts of the world to observe the spectacle at the appointed time and place with absolute conviction that it will take place on schedule. There is always the risk that the local weather will contrive to mar the view, but that the eclipse will occur is in no doubt whatsoever.

What is behind the incredible precision of such a prediction? How can we know so well the future spatial relationships between earth, moon, and sun? How can we be so certain exactly when, where, and for how long the moon will blot out the blinding light of the sun? Or, to take a simpler example, how do we know exactly where the planet Mars will be at any time in the future, so that an astronomer may, with perfect confidence and without so much as a glance at the heavens, aim his telescope at the predicted time and place and photograph Mars as easily as he does his daughter at her birthday party?

Generally we extrapolate the future course of the planets from the extensive data on their past motions that have been accumulated over the decades. The planets stay on schedule as does a train. Knowing where the tracks are laid and how fast a train travels, we can predict its arrival time at a given location. But unlike planets, trains are subject to accident, mechanical failure, and human error, so that we haven't nearly the faith in our railroad schedules that we do in our eclipses. What guarantees the near-perfect regularity of the motion of the planets is the law of gravity and Newton's laws of motion. Barring any cataclysmic astronomical events, the motion of a planet is completely determined by Newton's simple and elegant principles. If we know where Mars is at a given time, and how fast and in what direction it is traveling, Newton's method enables us to foretell the future course of Mars's motion, and even to specify its entire past behavior as well. We can, in fact, predict the future motion of any body, anywhere in the solar system, by applying to Newton's general laws the mathematical methods of the calculus.

Calculus retains a magical aura for me. In high school, I had a friend who shared my interest in science. We excitedly contemplated the secrets of science and mathematics to be revealed to us in our future college studies. Of special wonder were the unknown mysteries of *the* calculus. Usage in those days demanded that the definite article always accompany the word *calculus*, and I note with a certain nostalgic sadness that this seems no longer to be the case. We already knew something of algebra and trigonometry, but when should we become masters of *the* calculus? The intriguing use of the definite article heightened our awe and expectations. I was finally initiated into the mysteries of the calculus in my first college year and was relieved to discover that it wasn't half as bad as I'd feared. In fact I found it enjoyable.

(Completing integral calculus in summer school, I often delighted in sketching area-integration problems in the sands of Rockaway Beach.) But even after the novelty wore off, I continued to associate a certain magical quality with the techniques of calculus. Somehow, computing derivatives and integrals is never quite as straightforward an operation as addition and multiplication. It does become automatic with time, but whenever I stop to think about it . . .

In the seventeenth century, Newton and Leibniz independently invented the methods that today we call the calculus. Newton did this in conjunction with his work on planetary trajectories. When Newton's laws are applied to a planet, we obtain a mathematical expression for its acceleration. This is a measure of how rapidly the velocity of a moving object changes, just as velocity is a measure of how rapidly the object's position changes. When a car is moving at a fixed speed, its travel distance is increasing at a constant rate. When a car accelerates, its speed changes as well, and its distance increases at an ever-growing rate. Acceleration implies a varying velocity. What calculus does is to provide us with a mathematical means of computing a planet's trajectory or position from a knowledge of its acceleration. One passes, by well-prescribed steps, from acceleration to velocity to trajectory.

Alternatively, one may go from the trajectory to the acceleration. The techniques by which we go from trajectory to acceleration and vice versa are inverse mathematical operations like multiplication and division. Calculus enables us to leapfrog back and forth between acceleration, velocity, and trajectory.

We can understand this better by using an arithmetic analogy: Suppose we start with the operation "multiplication by 2" and its inverse operation, "division by 2." By successively multiplying and dividing the number 32 by 2, we can generate all the *powers* of 2 (2, 4, 8, 16, 32, 64, 128, etc.); all the reciprocals of these powers ($^1/_2$, $^1/_4$, $^1/_8$, $^1/_{16}$, etc.); and the number 1. These numbers and only these will be generated. But we could equally well begin with 16 or 128 or $^1/_8$ and produce the same family of numbers. With this operation and its inverse (x 2 and ÷ 2), the whole family of numbers based on 2 is generated from any one of its members. The same is true in using the methods of the calculus to generate acceleration from the velocity, or the velocity from the trajectory, or vice versa. (There are other members in this family, but they were of no interest to Newton, nor are they to us.) Starting from the laws of motion, Newton was able to calculate the trajectory of a planet from its acceleration or reverse the whole procedure.

But are we actually predicting anything in this procedure or simply revealing our own ignorance? The calculus is not magic. It does not pull rabbits out of hats. If we obtain the trajectory from the acceleration, the information on the trajectory is already implicit in the acceleration. We are using the methods of the calculus to help our poor eyes see what should have been perfectly clear from the outset. A more visionary creature than a human being might not need the calculus in order to extricate the trajectory from the acceleration; that creature would simply *see* trajectory and acceleration as the same thing. Shall we raise to the status of a prediction that which is inevitably contained in our original measurement (of acceleration, in this case)? Would we brag about being able to *predict* the sixth power of 2 once we know the fifth? This amounts to the age-old criticism of deductive logic, that the conclusion is already hidden in the premises and all we disclose is our own blindness. Whether we use the calculus or deductive logic like a magnifying glass to see what's already there, we're hardly producing new information or predicting anything new.

At this point, I anticipate a barrage of objections coming from certain readers, and perhaps even a feeling of outrage. They might protest in the following words:

> Hold on. What a pompous argument this is! Maybe *you*, the author, have divine knowledge and vision, but the rest of us mere humans do not. Of course, the information about the solar system is *there* already, but we don't know what it is. It has been the job of science to discover and unearth this information, and it has been damn successful at it. For thousands of years, humans had almost no knowledge of the laws governing the universe. But since the time of Newton, we have made enormous strides. What we now know may not be too impressive on some absolute scale, but it's far more than what Newton's predecessors knew. And it has all been done in the last 300 years. A stupendous achievement!

> Now you come along, standing comfortably, if ungratefully, on the shoulders of the giants who have sweated and toiled to gain the precious facts and principles that we have. And with a brazen gesture, you would sweep away our hard-won scientific discoveries and dub them obvious from the start. Easy for you to say! Of course we are ignorant. But there lies the glory of science: to dispel human ignorance through the painstaking discovery of facts and the laborious synthesizing of these facts into the laws of the physical universe. You are illustrating the power, not the weakness, of science.

Well, it seems that my antagonists are a bit literal about reality. Few philosophers would claim nowadays that the facts of nature are simply there in an objective external world and that the task of the scientist is to discover and interpret them. It is generally conceded that facts and discoveries are theory-laden and that they contain a subjective element. The case has been well argued by Poincaré, Polanyi, Kuhn, Holton, and Brown,[1] among others. However, I shall, at least, make one simple rebuttal.

My counterargument boils down to a question: On what basis have we chosen the definitions for acceleration and velocity? This may seem a strange question. We are accustomed to thinking that these are given quantities, highly suitable for describing motion. They seem obvious. What more natural way could there be to measure speed than by dividing the distance traveled by the time elapsed?

But how did we know in the early days of kinematics (the mathematical description of motion) that distance divided by time was more appropriate, for instance, than distance multiplied by time, or, to use a less preposterous sounding example, distance divided by the square of the time or even a logarithmic function of distance divided by time?[2]

The answer is that the quantities and the theories were invented in such a way that they fit together. There is no *a priori* reason why distance divided by time should be chosen as a measure of velocity. It is insufficient to say that it quantifies our vague subjective notions of speed, since many other quantities would do the same. The commonly accepted definition for velocity is used because it can be linked up in a deterministic way with corresponding definitions of acceleration and position in space.

This causal link was formalized by Isaac Newton in his application of the calculus to kinematics. In other words, the position, velocity, and acceleration of a moving body, which we take so much for granted today, were neither ordained for us nor discovered in nature. They were invented or adopted, either consciously or unconsciously, so as to fit a concomitant scheme of invented causal laws. The founders of modern physical science, including many unsung heroes in the Middle Ages and the Renaissance, constructed throuh their combined creative genius, a self-consistent system of quantities and mathematical laws employing those quantities. This system possesses certain internal connections which allow for prediction, i.e., for getting from one part of the system to another in a determined fashion.

It's all a little like inventing a game. If you imagine yourself inventing chess, you have to conceive of the board, the chesspieces with their moves, and the rules of play in close conjunction with each other. You could not get very far trying to use Monopoly rules and pool balls on a chessboard. The whole thing has to fit harmoniously together with efficiency and consistency that characterize a challenging, exciting, elegant game. A chessboard can also be used to play checkers. And checkers can be used to play backgammon. But this variability illustrates that each game, however unique, is one possibility among many. Each has its appeal and its devotees. Monopoly may attract those who seek riches and enjoy the game of chance; chess, those who would attain power through logic and strategy. Different games serve different moods and suit different tastes. No game can do all things for all people. Neither can science.

The things about science that make it useful to us and that make us appreciate it—predictability, objectivity, self-consistency, generality—do not exist in some external, independent reality. They are part of our experience and interpretation of the world, of our consciousness and values, of our game of meaning. I see Newton's monumental achievement as a mental creation, a humanly conceived world system, incorporating self-consistency and causal order, which pleases the human mind and helps to allay our fears of a chaotic universe. His is as much a work of art as it is of science.

To protest that Newton's conception is supported by countless observations of the physical universe is no argument, for my point is that the conception or theory and the observed quantities are created in parallel so as to corroborate each other (not necessarily without a struggle, and not in any overtly conscious way). Furthermore, the quantities themselves are based on a definition and procedures of measurement that are fundamentally subjective, as we saw in the last chapter.

In designing a railroad, the choice of locations for the stations and the tracks linking them is made *in conjunction* with the notion of a timetable, so that the whole thing will form a closed, predictable system that people can rely on and come to believe in. It makes no difference whether the designer begins with the tracks and stations or with the timetable. He holds in mind a conception of a self-consistent system, and he will choose among many possibilities until he invents a system that does what he wants. Afterward, if you point out that his railroad will not get you to your aunt's farmhouse or even predict how long it will take, he must concede that his railroad is not, after all, a total transportation system. It was not designed to do everything.

Neither was science!

Science is designed to deal with physical reality. But it is not alone in this realm. There are many systems of thought and belief, for example, of a metaphysical or religious nature, that have given countless people profound insight into the nature of physical reality. None of these can claim to predict and control nature as well as science does. But to choose science over other systems on the basis of its abilities to predict and control is a value judgment. It does not guarantee objectivity, only utility. Science is a value-construct, created by the human mind. But I do not advocate that we simply take the world as a subjective illusion, throw up our hands in despair, and hope for the best. Rather, it will be the burden of the latter part of this book to consider the meaning of the shared human participation in the creation of our reality-construct and to suggest taking fuller responsibility for it.

As soon as one admits the integral role played by the human

mind in science, psychological interpretation becomes essential. Simply to concede the creative and subjective nature of physical science, even of its most objective aspect, measurement, is not enough. It is necessary, in fact imperative, to understand how and why we perform this act of creation as we do. What are our underlying motives, beliefs, and fears? If we are to avoid using science to make a nightmare of human existence, we must make the act of creation as conscious as possible.

The attempt to interpret and understand human acts and thoughts in psychological terms is usually considered the province of the novelist, the psychoanalyst, and the philosopher. In recent times, a new school of psychohistorians has produced some impressive examples of how useful the psychological approach is in supplementing, and even enlarging, our understanding of the course of human events. Historians, philosophers and sociologists of science, like G. Holton, N. R. Hanson, J. Bronowski, E. Grant, E. A. Burtt, J. R. Ravetz, and O. Barfield, deal with the psychology of scientists. There is also the brilliant linguistic research of B. L. Whorf in elucidating the American Indian view of reality. These trends deserve encouragement. But I think we need a more personal kind of account of the thought of individual scientists. I do not mean that scientists should be publicly psychoanalyzed. What we want are probing studies of the metaphysical assumptions and beliefs that form the structure of scientific intellect. There already exist some attempts at this. Henri Poincaré was quite successful in revealing his own creative processes; Albert Einstein, somewhat less so. But, the standard practice among scientists of purging their publications of any trace of personal or individual motivations or other subjective considerations needs to be modified. We must eventually understand, for example, what motivates Kuhn's paradigms and Holton's themes. Shall we ever be able to control and guide the deep motives of science? It is the unrecognized and unconscious human element in scientific creativity that is posing a threat to humankind today, rather than the products of science and technology. It is we who create this world we live in, and we had better understand exactly what it is we're doing. Even railroad builders admit they're out to make a profit.

Generality and Generation

What I have argued thus far is that the predictive power of physics was built in and guaranteed from the beginning. Through a selective creation of quantities and laws, a self-contained system was constructed which gives answers on its own terms and arbitrarily rules out what it cannot deal with as nonquantitative or non-physical. When we apply Newton's method to Mars, for example, we begin with a knowledge of the trajectory of the planet (Kepler's laws), which is described in terms of operational definitions of space and time. These measurements, in turn, represent subjective value judgments, not objective reality. From these we contrive the causal, deterministic system of Newton's laws of motion and gravity, which inevitably enables us to forecast the already-known orbit of Mars.

Any symbolic discussion of Mars as the mythological god of war, for example, or as an archetype for the male principle is excluded from physical science on the grounds that such matters are nonquantitative and irrelevant, or, more rudely, are occult gibberish. The success of physics is based on the presumed wisdom of having long ago eliminated such unmanageable, nonquantitative matters from consideration. Physicists selectively narrowed their interests down to a group of questions and concepts they could handle, and, as time went on, they continued to justify their choice on the basis of ever greater triumphs. In retrospect, however, this favoring of quantification and causal determinism appears to be an aesthetic and subjective choice. We have infused the system of physical science with value and meaning for reasons hidden even from ourselves and have concealed our efforts behind the camouflage of objectivity.

I shall turn now to the final bastion of objectivity in physics: the generality of its laws. At this point, some readers might concede that predicting the orbit of Mars is contrived, but still they would argue that the same procedure works for a Jupiter or for the discovery of a Pluto. It even applies to the motion of objects in the sublunary realm, right here on the surface of the earth. Surely all of this cannot be part of a conspiracy, not even an unconscious one. There must be something real and objective about physical laws that are designed to work in one case and which later turn out to apply to many others. All bodies in the universe obey the same law of gravity and follow the same laws of motion. There has to be some new, uncontrived knowledge here. The objectivity of Newton's laws lies in their universality.

To answer this serious and important objection, I shall consider briefly some issues in the philosophy of science. I shall tread some ground already well cultivated by others and draw on the fruits of their labor. Henri Poincaré, in his book *The Foundations of Science*, tells us that mathematical truth is really a matter of convention or definition, *chosen* so as to avoid contradiction. Now Poincaré is talking about mathematics, but what he says could apply as well to Newton's method for developing a set of laws to predict planetary orbits. If in composing a set of laws or statements, Newton was careful to avoid contradictions, then he was ensuring internal consistency in the laws and the possibility of making predictions with them. Opting for consistency is a clear choice, one of convenience and utility no doubt, but a choice nonetheless. Our earlier discussion about predictability—about how one chooses a definition of velocity to blend and harmonize with the laws of motion so that they may be used to forecast the trajectory of Mars, for example—illustrates how the element of convenience enters into physical theories and suggests how Poincaré's notions apply to physics as well as mathematics.

Poincaré points out that we cannot claim the truth of one kind of geometry and the falsity of another. We cannot say that flat or Euclidean geometry is true, while spherical or Riemannian geometry is false. Space may be considered flat or curved under different circumstances. Which geometry a mathematician chooses to use is a matter of convention or convenience. Each serves a different purpose. To survey a plot of land for a building, flat geometry is the most convenient, whereas for navigating over large portions of the earth's surface, spherical geometry is preferable. One can navigate with flat geometry, however, as Mercator projections of the earth illustrate. Viewing the earth as a planet in three-dimensional space, it may be treated as a sphere, surrounded by a gravitational field, embedded in a flat space, or, alternatively, as a kind of curvature or "dimple" in space itself, so that the earth, together with its gravitational field, is described

We cannot say that flat geometry is true, while spherical geometry is false.

with the curved geometry of Einstein's general relativity theory. Poincaré's point is that mathematical truth is not based in reality at all but is a matter of convenience.

The situation is not all that different in physics. Choosing one theory over another is often a matter of convenience. I might, for instance, select a simple, approximate theory in order to calculate the spectrum of colored light emitted by a hydrogen atom, all the while recognizing that the more complex theory is more exact. But to verify that one theory is more exact than another, I should have to appeal to the objectivity of measurement, which I have already called into question. Thomas Kuhn, Michael Polanyi, and Gerald Holton have already demonstrated that much more than facts and measurement is involved in the choice of one theory over another, whether for the purposes of calculation or in the adoption of a whole new scientific paradigm. The seventeenth-century data on planetary orbits was not precise enough to distinguish between Ptolemy's epicycles and Kepler's ellipses. (In fact, one could still use epicycles today.) Kepler, himself, resisted his own ellipses for many years because they seemed to displace the sun from the central role in the solar system that Greek and countless earlier cultures had assumed it must play. Only when Kepler recognized in the sun's location at the *focus* of the ellipse a centrality analogous to the sun's central position in Greek thought was he able to accept his own elliptical paths of the planets and to publish his work.

Poincaré goes on to explain that "facts" are chosen according to a hierarchy, based on certain aesthetic criteria. We favor facts that reveal a certain simplicity and elegance and which seem to have the *highest frequency of recurrence*. Words like *simplicity*, *symmetry*, and *economy*, used to describe scientific theories, belie the value judgments behind them. They are discussed by Michael Polanyi in *Personal Knowledge*, where he says such words are "stretched far beyond their usual scope, so as to include the much deeper qualities which make the scientists rejoice in a vision like that of relativity."[3] Both Poincaré and Polanyi characterize this harmony-seeking inner sense, which guides our choice of facts, as an apprehension of a kind of rationality. It is the *mutual* appreciation of rational harmony, which we share with other thinking beings, that produces our sense of objective reality. Robert Pirsig in *Zen and the Art of Motorcycle Maintenance* puts it very well:

> What guarantees the objectivity of the world in which we live is that this world is common to us with other thinking beings. Through the communications that we have with other men we receive from them ready-made harmonious reasonings. We know that these reasonings do not come from us and at the same time we recognize in them, *because of their harmony*, the work of reasonable beings like ourselves. And as these reasonings appear to fit the world of our sensations, we think we may infer that these reasonable beings have seen the same thing as we; thus it is that we know we haven't been dreaming. It is this harmony, this *quality* if you will, that is the sole basis for the only known reality we can ever know.[4]

"These reasonings do not come from us," that is to say, they do not originate *in* us. They do not come from nature either, but they do come *through* us as a subjective experience. We participate in the creative process, even if we are unaware of it.

To understand how this sense of harmony relates to the generality of physical law, we must inquire into our sense of order and rationality. Poincaré speaks of ordered patterns as those that may be extended to unanticipated cases; Polanyi, as those that are highly improbable on the basis of human judgment. Here is the key—the inner meaning of order and harmony, of causality and determinism is their contradiction of chaos. It is the unexpected departure from randomness that we recognize as order and harmony, that surprises and delights us. Poincaré's hierarchy is a structure of human critical values, based on an antagonism to chaos and chance. It is we, not nature, who abhor the vacuum.[5]

The more unlikely a pattern is, the more orderly it appears to us. A snowflake fascinates us, for we see in it an exquisite pattern too improbable to believe. Where in the uniform and amorphous nature of water is hidden the secret symmetry of the snowlfake? The modern theory of matter begs the question. It explains the symmetries of crystals in terms of the arrays and lattices that are the spatial configurations formed by attracting atoms. These explanations in turn depend on the dimensional and metric properties of space and on the perfect congruence of identical atoms. Thus the improbable patterns of matter are deduced from the far more improbable assumption that like atoms are absolutely identical.[6]

It is the novelty and unlikelihood of order in the midst of chaos that attracts us. The planets were quickly singled out by early humans for their divergence from the uniform and random pattern of the background stars. (Even the stars were grouped into constellations and given meaning and value.) Thousands of years and untold human effort were spent in apprehending the order of the planetary system. The crowning glory of that effort was Newton's great synthesis. That Mars should obey a deterministic law reflects a high degree of order. That Jupiter should also obey one is more tidy still. But how much more harmonious and improbable that they should obey the same law.

Perhaps, my reader, you pause at this point in disbelief. You wish to ask whether I seriously claim that Newton, Kepler, and company, by some heroic, yet unconscious act, willed the great planets into similar orbits and to follow the same universal laws. Well, I cannot deny that I am suggesting a kind of collaboration between consciousness and matter. But I must also object that you are really begging the question by assuming what I cannot assume—that there is an independent, external, objective world out there, uninfluenced by human consciousness. From within an assumed framework of a mind/body dichotomy, I am hard pressed to envision or explain our participatory creativity or how we invoke order. Which came first, the chicken or the egg? Try asking an egg in the very process of becoming a chicken. In some higher state of awareness, in which all time is telescoped into the eternal now, the chicken and egg question becomes meaningless. Where past, present, and future are the same, chicken and egg are one. Where mind and matter are the same, creator and created are one. We cannot assume dichotomies when it is dichotomies we are calling into question.

But we can at least recognize the essential role that human discrimination and insight play in apprehending and articulating an element common to the motion of heavenly and earthbound bodies. Aristotle's earth/heaven dichotomy is motivated in part by the human desire to protect the order and harmony of the skies from the chaos of our mundane world. In modern times, humans could allow Newton to lift Aristotle's age-old barrier because they thereby regulated the downward flow of order, rather than permitting the dreaded upsurge of chaos. After Newton, harmony reigned on earth as well as in heaven.

Poincaré says that in selecting facts in accordance with our instinct for harmony and order, we seek out the large, the distant, and the microscopic. It is easier to discern what is exceptional and unlikely in what is remote. This is precisely what Newton did. He perceived the underlying order in the behavior of nearby objects on earth by studying the distant planets. What could be more unlikely than billiard balls and planets obeying the same laws? What more universal? Scientists do the same when they "simplify" a problem, reduce it to its "essential" elements, and seek some improbable connection with distant problems. What is this process of seeking out and eliminating, but an act of creation guided by human values and significance? It is we who cherish and create order by removing in our minds the extraneous and insignificant.

In similar fashion, the sculptor unlocks and informs his or her own concept of the beautiful by chiseling away the unwanted, unmeaning stone. The work of art, improbable, pregnant with meaning, emerges from the uniform, unexceptional rock. The artist makes a deliberate, conscious choice. It is an act of creation, not discovery. Guided by aesthetic intuition, the artist makes manifest this one statuary image from among the myriad potential images contained in the stone. Are there not as many physical theories, even physical realities, as there are works of sculpture contained in a block of marble? Do we not create them? And is it not possible that what we call generality is the reflection of our desire to value and glorify order and harmony and to ward off chaos?

Chapter 3.

Space to Let

The Cardinal Metaphors

In this chapter, we begin the major task of this book: to explore the four foundation concepts of physics—space, time, matter, and number—and their relationship to human consciousness. Because I believe that they *are* intimately related to consciousness and are guaranteed of no objective, external status by physics, I prefer to call them metaphors, in the sense that I used that term at the beginning of the introduction, as an act of consciousness that borders on the very creation of things. Thus I emphasize that they are creations of the mind.

In other words, metaphor suggests more than an analogy, for in a metaphorical comparison, a new quality or connection is disclosed in the things compared that was not previously apparent. Metaphor implies the creation of an idea or symbol, which not only stands for something else but, in fact, stands alone as a new evocation of meaning. Physics is a metaphor because it is a representation of a part of human experience (the so-called physical world) which breathes life and value into that experience and creates the meaning which that experience has for us. When we use a concept like space in physics, we are constructing an image, which despite its abstract character and quantitative description, evokes and summarizes what our being and existence feel like to us. Our spatial metaphor tells us what it is like to exist as an entity that is separate from, and yet related to, other things. Our experiences of location, distance, connection, and of self and other gain meaning and coherence through their spatial representation. We see ourselves located at a point in space, separated from other things by the distances of space, connected to them through light and sound which travel across space. What might otherwise be some kind of meaningless, inchoate experience of our own undifferentiated existence is transformed through the metaphor of space into an awareness of differentiation, diversity, separateness. Space organizes and gives meaning to our tangled, amalgamated experiences through its attributes of place and distance. Space has by now become so synonymous with all these experiences that we cannot see it as a representation of them, nor can we say clearly what is standing for what.

Space is only one of several important metaphors used in physics. We shall now examine physics as metaphor by exploring its four major constructs, space, time, matter, and number. Because these constructs are fundamental to our modern scientific conception of the cosmos, because they are the starting point for quantification, and because they offer a basis for wholeness, I call them cardinal metaphors. I see space, time, matter, and number as possibly the deepest expressions of the present state of our consciousness. They are a reflection of, and are reflected in, our language, imagery, thought, and belief. We shall therefore look to myth, dreams, imagination, and the arts in our reconnoitering of the cardinal metaphors. We shall seek their subjective aspects—their origins in consciousness, the motivations behind them, the meanings they have for us—and explore their continuity with the more objective notions we have about them.

It is no easy task to open ourselves to alternative ideas of space and time. We are all unknowingly indoctrinated into the religion of physicality. This is especially true within the study of physics, for it is almost impossible to be exposed to so profound and beautiful a subject without acquiring its habits of thought and succumbing to the spell of its power and success. But the problem extends far beyond the study of physics. It is not simply a matter of the modern habitual acceptance of television, computers, atomic power, and rocket travel. We recognize these as triumphs of science and technology and accept their evils along with their benefits. This, too, is a kind of hypnosis. But what I am referring to is a much subtler matter. It is the unconscious assumptions we make about the world which have largely been determined and reinforced by the development of scientific thought in the last few hundred years. It is in the texture of our consciousness: how we experience and view the very chairs we sit on and all the other solid objects we see and use; how we conceive of space, of our movement through it, and of the motion of external bodies; how we picture the dimension of depth and utilize other perspective notions; how we experience the passage of time; above all, how we feel ourselves to be isolated, physico-biological entities governed by the laws of matter, with our life and consciousness soon to be explained away by an edict from the biochemists and cell biologists.

You may think it strange to open such matters up for questioning. How else can one conceive of chairs and objects? Are they not solid? Do they not support us? You bang on the table before you (at least, I often do to make this point) and demonstrate the irrefutable solidity and substantiality of matter. What can be said in the face of such concrete evidence? Well, these are not easy questions to deal with. It is no simple matter to dislodge experiences and assumptions that are imbibed with our mothers' milk, and to conjure up alternatives to them.

A most impressive and convincing assault on our traditional view is Owen Barfield's seminal book, *Saving the Appearances*. Barfield argues that reality is a "collective representation," projected by an evolving human consciousness. But there can be no final proof of the reality or ideality of the world. Rather, we reach our convictions on the issue after deeply pondering matters in the physical world and our interpretations of them. It is a major burden of this book to help its readers reconsider their conviction or, at least, to help them realize that their views of space and time are matters of conviction. I shall try to present some alternative viewpoints, but two things are worth pointing out right now. First, science itself no longer views matter as substance. Atoms are not hard little balls. They are not even vague clouds or waves of probability as some texts are fond of saying. (Probability as such does not exist in and occupy space. Nor can probability itself be measured.) They are fundamentally unpicturable in the spatio-temporal terms we apply to gross matter. Notions of solidity in physics ultimately rest on such hypotheses as the Pauli Exclusion Principle (a generalization of the idea that two things cannot occupy the same space at the same time) or the existence of repulsive forces, the very notion of which already conceals within it the idea of extension or occupancy of space, and which therefore begs the question of solidity. We shall return to this important point in the chapter on matter.

Second, our concepts about space and time are held captive by our own prejudices. Space, for instance, is such a pervasive notion that it is all but impossible for us to conceive of existence without it. Or, to take a more comprehensible example, our knowledge of the existence of light frequencies just beyond the range of the visible spectrum (ultraviolet and infrared) does not enable us to see such colors or even visualize them in our mind's eye. Alternative symbols will be plentiful enough, once we have stopped idolizing our current ones, or better yet, stopped idolizing any at all.

The religion of physicality depends on many supporting sects. The strongest of these is physical science itself. That science is a religion and scientists its priests is often difficult to appreciate. If you are a layperson, you may believe in science and may worship privately at its altar, but for the proper meaning and interpretation of events and phenomena, you must turn to the priest for guidance. The physicist-priest is carefully schooled in the ways of science and in the administration of its sacraments through years of training, practice, and indoctrination by the elder members of the sect. When you come before a priest with a quest for understanding, you begin by confessing your ignorance. This state of ignorance is original and pervasive. Although compassionate, the priest knows that there is a limit to what the laity may receive. You cannot become fully enlightened. The wonders of science are there for all to see, but its deep mysteries remain the province of the priests.

A defender of physics may protest that this is farfetched, a gross and unfair exaggeration. Science, unlike religion, is based on reason and facts. And anyone with the will and talent can become a scientist and understand modern physics. But did not the medieval Roman Catholic Church always make the same claims? Were not its acts and dogmas endlessly justified by complex but eminently reasonable arguments? Was there not always the most painstaking and exhaustive investigation of the facts surrounding miraculous events? Scripture could be understood by anyone willing to spend the necessary years in learning Latin, ancient Greek, and Hebrew, just as modern theories of science can be learned by those with the necessary background in mathematics. No one is excluded.

What I am saying is that science and religion both go wrong when they become dogmatic and idolatrous. They have a common subjective origin as creations, designed to ameliorate the human condition. Science, art, and religion all serve an essential human purpose and proceed from subjective, psychological motives. More of this later. We turn now to some alternative space metaphors.

The Space Theater

Imagine a visit to the theater in a carnival of space. In the auditorium you find yourself flying over the skyscrapers of New York. They erupt vertically toward you, separated by a rectangular grid of dizzying narrow valleys. It's like the opening of the film *West Side Story*, seen in 3-D wraparound. The buildings, like great rhomboidal crystals, sway ponderously and silently below you as your changing perspective has them tilt from the vertical, now this way, now that. The view of the rooftops, towers, and spires alternates with that of the streets far below, rapidly telescoping your perception of depth and height—up, down, in, out, high, low. Your gaze falls deeply into the space below. You sense the frightening penetrability of space, its unfathomable depths, its infinite reaches.

Suddenly, you are over the great canyons—Grand, Bryce. The extremes of height and depth still appear, but now, the rectangular regularity of the skyscrapers is replaced by organic, irregular shapes—buttes, mesas, pinnacles, cliffs. Views of the Alps and Himalayas follow, bringing with them a gradual heightening of depth perception and diminishing sense of proportion and size. Confusion increases as scenes alternate and overlap. Are these mountins you see or stalagmites? Are they microscopic crystalline growths or perhaps greatly enlarged organic tissue structures? Are you looking upward into the pendant vines and mosses of a tropical hanging garden, or down among the towering pines and redwoods of a coastal rain forest? Is that a morning mist permeating the air between the walls of a ravine or are you beneath the murky waters of the ocean, peering into an undersea gorge? What are those swaying tubes approaching you, the cilia of a magnified paramecium, the tendrils of a Venus flytrap, the tentacles of an attacking octopus? Do you really feel something? Is this a visual hallucination, a tactile illusion, or reality? What is that gaping abyss, the jaws of hell, the maw of a whirlpool, a bottomless black hole?

You decide you've had enough. You get up and head for the exit. But the door seems to be a projection on the wall like the views above you. There doesn't seem to be an exit. This is quite a theater, you muse. You decide to sit on the floor and wait for the "show" to end. You're relieved to find that the floor is solid and not a projection. You won't be returning here too soon, you think. You close your eyes to shut out the disturbing visions, but nothing changes: you still see them! What's going on? Have my eyelids become transparent? Are they projecting pictures inside my head? Must be some newfangled mental projector. What ever happened to the nice simple, clean space we all knew and loved so well, you wonder. It always knew its place and stayed there—outside of me. But the space in this theater is all somehow connected. There's no clear inside and outside. And it seems to be more than just space. It seems to be sensations and feelings as well.

The view, or experience, now is strangely organic. It started out something like a jungle, but the twisting vines and branches are partly transparent and appear hollow. Fluids and bubbles seem to percolate through them as they reveal themselves as connective tubules that branch into ever finer filaments like the capillaries of the circulatory system. The fluids and vapors they carry are more than physical. They are thoughts and feelings, and more strangely, abstract ideas, symbols, and meanings. Confused by this tangle of currents, emotions, experiences, and thoughts which are at the same time physical, psychological, and intellectual, you start to rise once more to escape from it all. You discover, to your alarm, that the tubules and currents penetrate your own body, that there's no clear boundary between your inside and the outside! You fear that you're trapped in this strange tangled web, but you find that you can, in fact, move (if you can call it that). You no longer seem to have a body in the normal sense of the word. Rather, you are like a vortex or a pattern of concentration among all the flows and channels. To your amazement, it's all rather pleasant. You're not so much tangled up in things as you are connected to them. You are aware of thoughts, feelings, and experiences that are not your own, not necessarily even those of other people. When you wish to move, you pass effortlessly like a wave through the connective matrix. It's more like a thought moving through a mind than a body moving in space. Your former experience of space as an empty, geometrical void in which things have an existence as isolated entities, separated from each other by distance, has been transformed. You now experience a realm filled with meaning and wisdom, and in which things blend and participate in each other's being and significance. The projective, perspectival geometry of the artist and physicist has become the participatory connectivity of the alchemist and seer. Space is mind. Here meaning and information are not transported *across* space and *in* time. Instead, they are shared and omnipresent, for the fundamental relations are symbolic, rather than spatial. It is meaning which connects things, not distance which separates them. Causality is enlarged to synchronicity. This is a domain somewhere between the multiplicity of everyday consciousness and an ultimate unity.

Space a la Mode

We tend to think of space as rigid, fixed, and external. But our experience of space varies with circumstances and history. Earlier people did not see and feel space as we do. Immanuel Kant said that space, time, and causality are conceptual and intuitive categories, constructed by the human mind in its response to matter. They are categories for organizing our sense perceptions and experiences. They are neither purely of the mind nor of the world, but a product of their interplay. Without the concepts of space, time, etc., the amount and complexity of the sense data we receive constantly through our eyes and ears would create an incoherent and confusing tangle of stimuli and mental images. Space, with its dimensionality, extension, and depth, provides room in which things may spread out and become separate from each other; it offers a structure for sorting out the bewildering impressions; it brings order out of chaos.

Do the sources of our sense impressions have an independent reality? Kant does not finally answer this question. He does imply the existence of an external reality: he refers to *things-in-themselves* which are the ultimate, but fundamentally unknowable sources of our perceptions. All we can ever know through our powers of observation and reasoning concerns the nature and behavior of these perceptions, and not of the *things-in-themselves* from which they are presumably derived. Thus the intellect can never be a source of information about the ultimate nature of reality (if any). And this has largely been the position of science, which seems too busy with what the intellect *can* do to be bothered about what it presumably cannot.

My own viewpoint is somewhat different. I do not draw any sharp distinction between some ultimate entity and our perception of it. Thus I cannot talk about our mental constructs as simply a response to some independent, albeit unknowable, reality. I think of a whole continuum from construct to object as a human creation, although not a capricious one. What I take as ultimate experiences are more psychological than physical, and I shall develop this theme later. But for now I must observe that the main problem is the framework in which this deilemma is conceived. If one begins by assuming a separation of subject and object or of mind and body, the boundary between them sooner or later becomes problematical. And such separations are so fundamental to our thinking that it is difficult to conceive of alternatives. Indeed, it is the very notion of separation and articulation that is at the root of our spatial metaphor.

When we picture space today, we tend to visualize an empty void without qualities. It has the fundamental property of *extension* from which we derive our ideas not only of the quantity and measurability of space but also of its capacity to be filled or occupied. It also has a characteristic continuity and dimensionality. None of this, however, adds up to anything very substantial; space is just about the most vacuous concept we have. We conceive it as the absence of everything: nothing but pure extension, the capacity or potentiality to be occupied by matter, yet not itself matter. If we look at one of the marvelous perspective drawings, say of DaVinci, which seems to capture so perfectly the character of a three-dimensional reality on a sheer two-dimensional surface, we immediately comprehend the laws of perspective which the artist so deftly employs. If we reflect on such matters, we are delighted that these laws were discovered in the Renaissance, and amazed that artists should have struggled for so many years prior to that time to portray space properly on flat planes. It seems incredible that the simple and obvious geometrical constructions, which we call the laws of perspective, should have escaped artists for so long. And medieval and premedieval painting will always appear quaint and naive to us, at least in its representation of space.

It is difficult for us to conceive of space as other than an empty void to which the laws of perspective can be applied, and this has led us to believe that there is something objective or absolute about our spatial notions. But the preeminence of our own view is not so difficult to challenge if we look at the art forms, belief patterns, and languages of societies remote from ours in time, place, or experience. For example, to the mind of medieval people, space did not have the cold, empty, geometric character that it has for us. We experience ourselves as an isolated, disconnected entity in a vast, empty void. But medieval people felt more a part of their surrounding environment. They felt a kind of extrasensory, but conscious, connection to the plants and animals around them, to the heavenly objects, to the very elements and minerals of the earth itself. We tend to dismiss such experiences as primitive or misguided. We explain them away as examples of animism and anthropomorphism. Or we reject them along with the gibberish of astrology and alchemy, which have been discredited by modern science. But, at the moment, we are challenging the assumptions of modern science and may not use them to reject other belief structures. We are obliged to take more seriously the notions of other civilizations. The sensing of human character and life in so-called inaminate objects by many primitive and even pre-Renaissance peoples should not be viewed as some kind of magical network of relationships existing within space as we know it today. Rather, it is part of the actual experience of space for these peoples. Suppose I, as a medieval man, diagnose myself as mercurial and I symbolize this through a *felt* connection to the planet Mercury or the element quicksilver in the earth's crust. Within the framework of my medieval mind, I do not picture all of this to myself as some mysterious influence of the planet Mercury acting at a distance across the vast reaches of empty space or as some chemical spell that quicksilver casts over my personality regardless of its distance and separation from me in the earth. These are not relationships *in* space to me; they *are* space. It is the sum of all the felt organic connections between my inner and outer worlds that I experience as space itself. Space is the synthesis of all my feelings of relatedness, connectivity, orientation. Owen Barfield summarized it beautifully:

> The background picture then (in the Middle Ages) was of man as a microcosm within the macrocosm. It is clear that he did not feel himself isolated by his skin from the world outside him to quite the same extent as we do. He was integrated or mortised into it, each different part of him being united to a different part of it by some invisible thread. In his relation to his environment, the man of the Middle Ages was rather less like an island, rather more like an embryo, than we are.[1]

But even such notions as relationship and environment conceal from us the more remote and unfamiliar wellspring of our modern spatial metaphor. A clue to the more organic view of space of so-called primitive societies may be gotten from the notion of the totem. Nowadays, we view animals and objects as different, distinct, separate from us, and so we find it hard to imagine or feel any kind of relation to them. At an earlier period, a person's relation to the totem was through ancestry or kinship. Such a blood relationship was felt inwardly, as a kind of participation, and accepted as part of life and experience. In the very earliest times, the totem relationship was one of identity. No separation was felt between one's self and the totem. The original totem experience or relation seems to have evolved from identity to participation to separation, if we look upon our present attitudes about animals and objects as distinct from and unrelated to us as a kind of negative totemism. All of this hints at a totally different experience of space from our own. These relationships of kinship, participation, identity cannot be viewed as bridging the gaps of space; they are space. They must not be interpreted as geometrical and physical, but as organic, psychological, mythic, symbolic, and metaphorical. The Renaissance artist did not discover properties of space which had always been there for all to see; he began to construct and experience a new kind of space. He was far more an inventor than a discoverer.

Our modern notion of space is a compound metaphor that embodies all our concepts and experiences of separation, distinction, articulation, isolation, delimitation, division, differentiation, and identity. Fundamental to all of these is the idea of distance, for one cannot distinguish and identify overlapping things. One cannot articulate objects that cannot be resolved and distanced from one another. A most fundamental and universal property of space, embodied or implied in all physical law, is that two things cannot occupy the same place at the same time. This impenetrability of the point and the idea of extension are the essence of our spatial metaphor. The laws of perspective and of geometry for us are a codified summary of our normal experience of alienation, unique identity, and unrelatedness. It has all been abstracted, externalized, and synthesized into the cold, empty void we call space. But it is all our own doing and the result of idolizing our creation. Yet we continue to believe that space is simply there independent of us, and so it always was and always will remain.

The metaphor of space is our modern mechanism for avoiding the horrendous experience of oneness, of the chaos, of the ultimate state of unity to which the mystic seers and philosophers of all ages have referred. We fear that state and picture it as one of undifferentiated dissolution and nonidentity. It seems to be the antithesis of space with its extension, articulation, and impenetrability. Our very fears are felt in spatial terms as we imagine being swallowed up, dissolved, and blended with all other things in some primal uniform glob. We cannot even conceive of these cataclysmic processes except as occurring *in* space and time. Such is the pervasive and ingrained nature of our spatial metaphor, which we have long idolized and believed to be without alternatives.

It might be objected that geometry was discovered long before the Renaissance and the Enlightenment by the ancient Egyptians and Greeks and that its techniques were applied to a distinctly spatial conception of the motion of heavenly objects, if not to painting. To answer this, one must turn to historians and philosophers of science,[2] beginning with Plato, to understand the role played by premodern geometrical models in "saving the appearances" of the phenomena. Geometry in its application to astronomy was used by the ancient Greeks, and roughly until the time of Copernicus, to reconcile the outward appearances of events and phenomena with their inner, concealed, true meaning. The irregular and chaotic external appearance of things was to be saved or reconciled with their inner harmony and order by means of an intermediate geometrical model or hypothesis. These hypotheses to save the appearances were frankly recognized as useful representations, not as truthful descriptions of reality. All of this was changed by the Copernican revolution, which stands as a major turning point in the evolution of the notion of hypothesis and of the Western view of reality. Since the time of Copernicus and Bacon, any collection of mathematical and geometrical statements, capable of saving *all* the appearances of the phenomena, have been given the status of physical law, accurate statements about reality, and not of mere hypotheses. It was for this reason that Newton made his famous disclaimer, "I do not construct hypotheses," (although the modern connotation of hypothesis as a tentative postulate subject to experimental verification has blurred Newton's intended meaning). Thus the development of geometry and its use as a construct or hypothesis prior to the modern era does not reveal to us how earlier peoples pictured and experienced space. It is likely, in fact, that it took humans 2000 years to begin taking seriously their geometrical models, to treat a representation as something physical. Yesterday's metaphor is today's reality. Modern space is a recent invention.

There is another aspect of our modern spatial metaphor worth considering for the light it can shed on our earlier discussion about counting. In chapter 1, the counting operation, in the process of measuring length, was traced back to the ability to articulate and identify the discrete elements of a set. We saw that an essential step in determining the length of a table was to make a one-to-one correspondence between the unit-length segments laid off along the edge of the table and the members of some discrete set, such as elephants or simply geometrical points. This in turn was seen to depend on the human ability to articulate, recognize, and identify the elements of the set. It now becomes clear that the notions of separation, discreteness, and articulation, so essential to counting and in turn to measurement, are among the fundamental characteristics of space as we conceive it. We cannot even count without something like our spatial metaphor. Thus our constructed space has precisely the properties needed in order to measure it. Again we can see the built-in circularity and self-consistency of our metaphor of reality.

Identification or *singling out* (a fine spatial metaphor, if ever there was one) presupposes an extended background of differentiable things, objects, points, what have you, that can be separated from each other. This is part of the design of our spatial metaphor. It spreads out and disentangles the otherwise continuous, chaotic experiences of what may well be a supraspatial or nonspatial world. The delimitation of objects and their isolation from one another is a created construct of the human mind or, at the very least, a necessary consequence of its peculiar method of sorting and filtering sense data á la Kant. Our ideas of extension and divisibility are synonymous with our conception and experience of space. Space *means* extension, distance, apartness, isolation. All these characteristics are mental constructs, along with the idea of a distinct object, and even of number itself. Space, measurement, and number are consequences of our fear of chaos.

Astrological Space

A very powerful spatial metaphor alternative to our own may be discovered through a sympathetic consideration of astrology. The unity and connectedness of all things is reflected in the astrological blending and equating of the inner and outer realms of consciousness and space. Astrology holds up a mirror to human consciousness. It reflects the medieval feeling of being imbedded in the cosmos.

If we try to understand astrology sympathetically from within the Greek and medieval forms of consciousness that nurtured it, we find that the underlying tenet of astrology—"as above, so below"—is decidedly inconsistent with the conception of an inner subjective world of mind and an outer objective one of reality, which is usually assumed as a matter of course today. Our idea of above and below or of outer and inner is that of two different spatial realms.

For the medieval astrologer, above and below refer not to different places but to different aspects of the same thing. There can be no above without a below. The two are connected—in fact, unified; and the many correspondences and felt relationships between them inform the study of astrology. Astrology is thus the explication of the connections that exist between the stars and humans, between two apparently different realms which are actually one. Since the experience of this unity is not ordinarily given to us, we find correspondences, echoes, hints of one realm in the other. In using the stars to study ourselves, we tacitly accept this unity and find our inner selves reflected in our experience of the world and its space.

Medieval consciousness did not feel so keenly as does ours that the mind and the rest of the organic space of the human body is bounded by our skin. We feel ourselves to be well defined and delimited from everything outside ourselves. What is inside is me, and what is outside is other. There is nothing in between and no overlap. If I think about it, there is some ambiguity when I eat. Some of my intake remains foreign and is expelled. The rest turns into me. I might have some trouble determining the exact instant the foreign substance becomes me, but, at some point, it is my living tissue which I can use to feel with or pump blood with or think with. When I conceptualize about my body in this way, it's almost as if my body isn't me either. I seem to be an agent using my body, which therefore becomes somehow other.

This gradual diminishing of the me is a characteristic feature of modern consciousness and space. As I think about myself in terms of the physical, chemical, and biological processes which science has been so successful in describing, I picture my consciousness or my self as occupying a shrinking region, somewhere inside my head (at least, that's where I seem to feel it). Where does that leave me and my inner space? Indeed, without the revolution in mind expansion of the 1960s and 1970s, the crushing weight of the arguments of modern science might well have convinced us by now that we don't exist at all, that only what is other is real, that consciousness and even life itself are ephemeral and illusory.

Astrological or medieval space, by contrast, has none of this abstract, lifeless character. It would not even have a purely spatial character, were we to experience it. Much of what a medieval person would think of as spatial, we today would call mental, emotional, or psychological. To the medieval mind, space, or let us say spatial relationships, comprises the felt connections among things. If a knight was called quixotic or mercurial, this was as much a spatial as a psychological statement (to use modern categories). The knight and the element were two different aspects of the same thing, which were connected in some way that was dimly perceived by the medieval mind *as* space. There was no separate, external, geometrical space in which to picture abstract relationships. Rather, there existed among things a web of organic and reflexive relationships, whose quality for the medieval mind was analogous to our sense of space.

In such a realm of organic connectedness, the medieval astrologer pondered the relationship of humans to the stars. He did not think in the terms that we might use of the influence of the planet Mercury on someone at the moment of his or her birth, projected across millions of miles of empty space. Rather, he recognized in the primal moment, when a newborn child drew its first breath of life, the stamp of a unique event impressed upon the whole cosmos and reflected in its every rhythm and pattern. He might equally well have read the child's essence and potential in other reflective and synonymous patterns—in the waves and currents of of the sea or the fluttering leaves in the forests or the elements of the earth or the stars in the sky.

Astrology does not concern itself, therefore, with cause and effect. It makes no more sense to say that Mercury has cast a spell on the newborn baby than it does to say that the baby has cast one on Mercury. It isn't that either one affects the other, but that they reflect each other. The whole configuration of earth and sky is a profound symbol for the child and for Mercury's momentary harmonies and relations to other heavenly bodies. In this important sense, astrology is antithetical to modern thought and science, which tend to suggest that the cosmos and its causal connections are without meaning or purpose. Our modern, meaningless, random universe would be inconceivable to the medieval astrologer. Meaning and wisdom are incorporated in astrological space, which is symbolic, organic, and synchronistic, rather than empty, geometrical, and causal. The spatial relation of Mercury to the child (in our modern sense of space) is of little importance in medieval astrology. A natal chart represents the organic and harmonic relations among the various astrological elements rather than the geometrical ones. It is the organic, reflective, symbolic relation that is of primary importance, and this connection is felt intuitively by the astrologer, as it was by ordinary people in the Middle Ages.

The medieval person felt connected to Mercury in much the same way as you feel connected, let us say, to your liver. The geometrical location of your liver scarcely begins to suggest its basic relationship to you. It is your liver's organic and functional relation to you that is really important. It purifies your blood, aids your digestion, and stores some of your energy. In turn, it is nourished symbiotically by the organs and systems of your body, which it serves. All of this happens in a smooth and functional way which cannot possibly be described adequately in spatial terms. Your liver functions holistically and purposively in concert with the rest of your body. You are an integral whole. Your liver is not really separable from you, spatially or in any other way. If someone tried to convince you that your liver is really somewhere far out in space and that it carries out its function by mysteriously raying its products to you across the empty miles, your incredulous reaction would not be very different from that of a fifteenth-century astrologer who had just been told

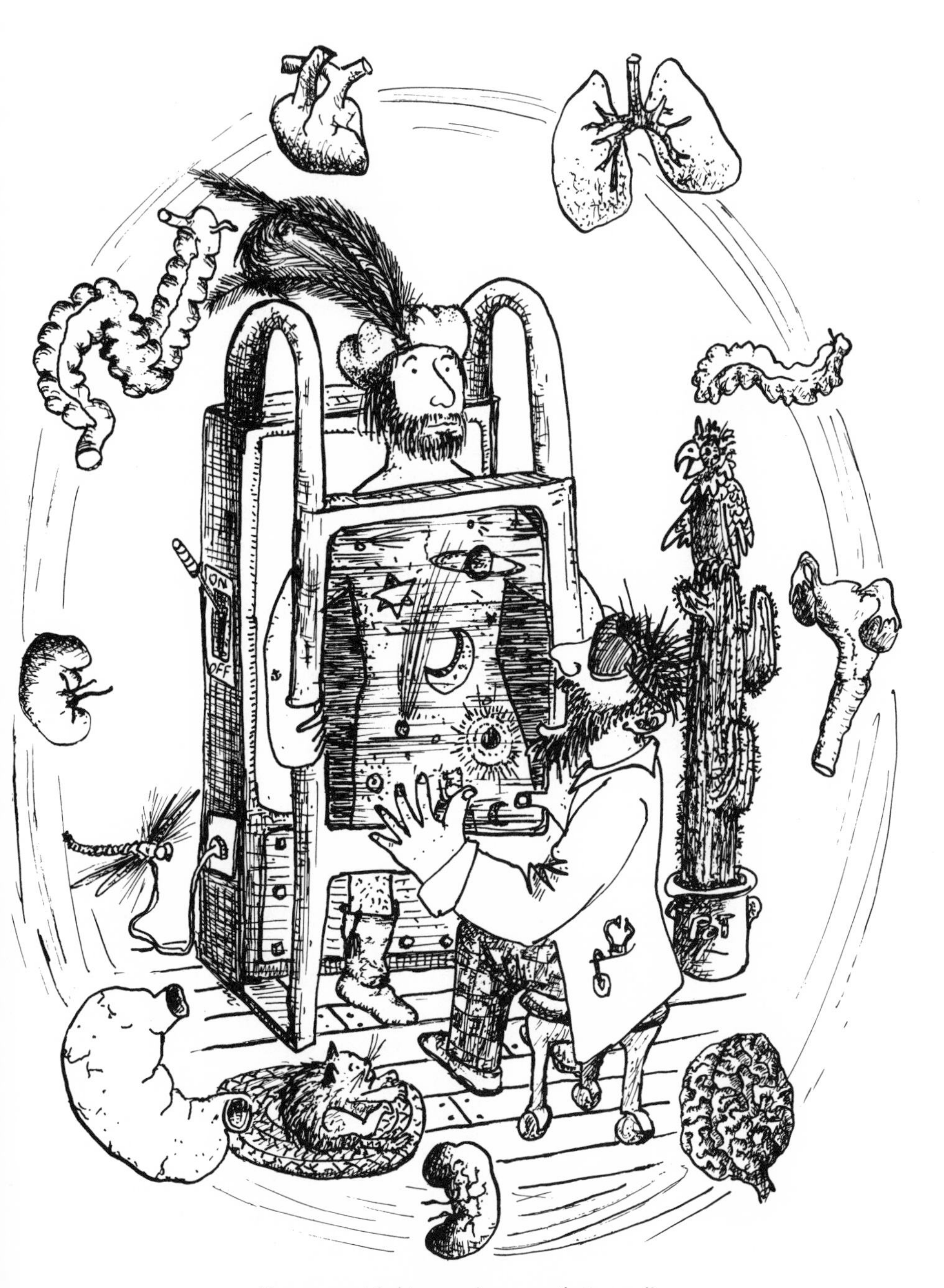

If someone tried to convince you that your liver
is really somewhere out in space . . .

that the planet Mercury is located millions of miles away in space and is not organically connected to him, nor in any way a part of him.

The modern distinctions between symbolic and realistic, metaphorical and literal, inner and outer, subjective and objective have little meaning for medieval consciousness. To medieval astrologers, alchemists, and artists, seeking the unity of all consciousness, life, and being, what possible sense could there be to a space that is abstract, external, and perspectival? They knew that space was a metaphor, a symbol for all the interrelationships and harmonies among the stars, the elements, and humans. In modern consciousness, mathematical laws of cause and effect in geometrical space are the emasculated vestiges of the connections we share with all things, vividly sensed by earlier consciousness.

Since there is no way to separate our knowledge of the world from our consciousness of it, changes in either world or mind must be understood as changes in both. World and mind are seemingly different, perhaps complementary, aspects of the same thing. Thus a different view of the world held by earlier peoples, one which to us is simpler and more naive than our own, is not a reflection of inferior knowledge, but of a different metaphoric expression of the world resulting from a different state of consciousness. The evolution of consciousness then follows from the inseparability of mind and matter and from the recognition that other people have experienced the world differently from ourselves. This requires that we accept a "primitive" world view as a serious, sympathetic, and accurate description of an earlier experience rather than as uninformed, superstitious, or inferior. If experience has changed and if consciousness is inseparable from reality, then there has been an evolution of consciousness/reality.

In modern times, our lost sense of synthesis or connection has become intellectualized as an assumption about reality (that it is separate and independent of our inner mental world and, in fact, subsumes that inner world, which is therefore not real. We conceive of space as an infinite,[3] empty, lifeless, cold,[4] dark, alien void. It is the blank, unfeeling stage on which matter plays out its aimless, random acts. It provides the merest, tiniest corner in which to harbor an insignificant speck of a planet, warmed by a second-rate star, on which by sheer accident, against impossible odds, life and finally consciousness have come to be. We see ourselves as living in a basically alien universe which offers us little succor or hope, and above all, no meaning or purpose.

Our modern space is the perfect metaphor for separation, extension, individuation, and alienation. We cannot even conceive of existence except in space, which then becomes the medium par excellence of existence. *To exist* is derived from the Latin verb meaning *to stand out*, and space is exactly what we stand out from. Space is the background from which we emerge or exist, in which we become an articulate, individuated, unique being. On one side of the coin is existence and uniqueness, on the other, alienation and isolation. Our spatial metaphor is thus intimately linked with our fears and apprehensions about life, death and survival.

The space of medieval consciousness, by contrast, is organic, connective, nurturing, human, intelligent, alive with meaning. It is a realm of wisdom and a storehouse of knowledge. Rather than space, it is place, home, environment. Like a womb to an embryo, it sustains, warms, and nurtures; it provides balms and lifelines; it has no clear-cut boundaries, no separation between inner and outer. Although less sharply defined, clean, and geometrical than our space, it contains things that we would not think of as spatial at all, things psychological, emotional, intuitive. One's feeling for others and for other living and inanimate things are included, so that the sense of medieval space incorporates love, appreciation, inspiration, belonging, kinship, and holiness.

In the medieval world, you felt somewhat less individual, but much more a vital part of things. You belonged to some great organism and functioned meaningfully and purposefully within it. The meaning and purpose might not be clear, but it was there all around you. You could feel it, sense it. Astrologers and alchemists sought it in the stars and the elements, whose connections to you were not *in* space, but *were* space. You were basically at home, supported, succored. Life might be difficult, but never foreign. Death might be frightening, but not annihilating.

We dismiss astrology, not so much for the reason that it does or does not work, but because of the anomalous kinds of explanations it would require in our current physical and geometrical theories (the latest "paradigm," as Kuhn[5] would say). Carl Jung uses the notion of *synchronicity* to provide some kind of explanation for the divinatory powers of the ancient *I Ching*[6] and of astrology. Synchronicity is the principle of noncausal links between coincident or symbolically related events. An example would be the cardinal I sighted outside my window while writing the paragraph on cardinality. Another is the correspondence between a prominent position of the planet Mercury and the birth of a child with dominant mercurial traits.

Synchronicity is no explanation at all in the accepted scientific sense. For an acausal principle is a contradiction in scientific terms. But rather than reject synchronicity for this reason, we must recognize its power to help us transcend our limitations. The real contradiction lies in the incompatibility of synchronous and causal consciousness or of astrological and physical space. And the incompatibility stems from our refusal to apprehend and accept the relative metaphorical status of all our constructs. As long as we insist that one kind of space is literal and objective, while another is metaphorical and mythic, we shall always be bound by the apparent contradictions between them. For these different views are complementary, and neither of them can give a complete picture. Human enlightenment is probably the result of the rare synthesis of all the many views, metaphors, constructs, and can never be completely experienced, understood, or communicated in terms of any one of them. To the extent that astrology is already a partial synthesis of some of our most meaningful metaphors, we do ourselves a grave disservice to ignore its wisdom.

Indeed, we do ourselves much harm in refusing to seriously study and contemplate many of the mythic and symbolic systems of early, primitive, and alternative cultures. Our chronic inability to come to grips with such matters as totemism, animism, reincarnation, extrasensory perception, divination, and a world of mythic creatures, archetypes, and gods and goddesses with human traits stems from our idolatry of physical and geometrical space. It is space which separates and distinguishes things from one another and then requires a causality to link them again. Implicit in Jung's principle of synchronicity is the idea that all things are reflections of each other: every pattern, every event is a microcosm of the whole universe. This holistic and fundamentally nonspatial notion has its roots in all ancient cultures. That all is derived from one—is One—is buried deep in human consciousness. If I can read my personality in the stars, it is because I and the stars are one and the same—are but different metaphors.

Before leaving the matter of astrology behind, there is one more metaphor I'd like to suggest with its help. The method of astrology provides for the construction of a natal chart for every human being. This chart is a symbolic map of the heavens at the time of birth of a person from which a gifted astrologer can read that person's basic character and destiny. (It is not all as fatalistic as it might sound, for astrology deals in potentialities and limitations which define the broad outlines of events but do not determine them absolutely.) Thus each personality and its destiny is roughly correlated with a birth or natal star configuration and its subsequent unfolding pattern. In other words, the entire collection of human personalities and life stories is an alternate manifestation or metaphor of the evolving events of our solar system as viewed from the earth. All the complex occurrences and relationships of human life and society have their counterpart in the spacetime events of astronomy. All human history is yet another spacetime metaphor.

Cartesian Space

Another transformed metaphor for space is one we all learn about in high school. It is the analytical geometry of René Descartes. Nowadays the brilliant legacy of Descartes is viewed with some misgiving. For as one of the founding fathers of the Enlightenment and the crucial mind/body dichotomy, he has bequeathed us a painfully sharp double-edged sword. His genius, which has cleanly cut through the dense web of primitive mystery, has also severed our felt connection to the universe. How ironic, then, that Descartes might provide a means to reunite the rent parts.

Analytical geometry is a correlation of algebra with geometry.

By means of a simple and elegant representation, the now famous graph-grid with its perpendicular "X" and "Y" axes, Descartes was able to show that any algebraic equation in two variables (x and y, for instance) is a precise description of a figure in plane geometry. For example, a circle of radius r is given by the equation $x^2 + y^2 = r^2$. In this equation, x represents the distance along the X-axis and similarly for y. Thus any figure in two-dimensional plane geometry can be transformed into a mathematical or logical statement, and vice versa. Furthermore, the whole technique can be carried over into three dimensions by using algebraic equations in three variables. And clearly, algebra and geometry may be viewed as transformations of each other. Thus a certain large collection of mathematical statements (those that describe all possible geometric figures) constitutes a perfect metaphor of space with all its inherent geometrical properties. And this metaphor is expressed in logical and symbolic terms, rather than in geometrical ones.

We have here a metaphor for space in totally nonspatial terms. It is a linguistic version of space, and it is perfect. For there is no information about space which is not contained in its algebraic metaphor. No poet could ever accomplish such a feat (at least not in any normal state of consciousness). It would be tantamount to expecting of Shakespeare's words alone a complete and perfect description of the man Hamlet.

Now it is true that algebra is not a natural medium of thought and communication for most people, but neither is Russian for most Americans or musical notation for most Eskimos. The point is that algebra is a form of language and logic which is treated very differently by the mind than spatial systems. Some people can deal more easily with logical concepts and others with geometrical ones. The talent for music or painting or language or mathematics may simply be the predilection for one particular space metaphor. It is entirely possible to view all systems of human thought and communication as fundamental metaphors of space, or, inversely, to see space as a projected metaphor for the whole world of the human mind. And this idea is traceable to the texture and thought of Plato's philosophy, with its tendency, as Owen Barfield puts it, "to experience as one what we now distinguish absolutely as 'mind,' on the one hand, and 'space' on the other."[7]

Descarte's brilliant synthesis of algebra and geometry gives us a revealing glimpse, even through the screen of its formal mathematical structure, of the ghost of a former consciousness in which there was no separate inner and outer world and the activities of the mind and of moving objects in space were somehow experienced synonymously. It was a view of humans as "surrounded by a cosmos or sphere of widsom"[8] rather than by empty space. Mind, thought, space, and motion were one and the same, and language reflected this amalgam.

The Tree of Man

By way of contemplating another transformed spatial metaphor, we shall consider some of the ideas of the philosopher, mystic, and mathematician P. D. Ouspensky. In his thought-provoking book *Tertium Organum*, Ouspensky tries to help us think about and visualize things outside our normal spacetime perceptions:

> Let us imagine some object, say *a book*, outside of time and space. What will this last mean? Were we to take the book out of time and space it would mean that *all books* which have existed, exist now, and will exist, *exist together*, i.e., occupy one and the same place and exist simultaneously, forming as it were *one book* which includes within itself the properties, characteristics and peculiarities of all books possible in the world. When we say simply, *a book*, we have in mind *something* possessing the common characteristic of all books—this is *a concept*. But that *book* about which we are talking now, possesses not only these common characteristics but the individual characteristics of all separate books . . .
>
> What is "man" out of space and time? He is all humanity, man as the "species"—*Homo Sapiens*, but at the same time possessing the characteristics, peculiarities and individual earmarks of *all* separate men. This is you, and I, and Julius Caesar and the conspirators who killed him, and the newsboy I pass every day—all kings, all slaves, all saints, all sinners—all taken together, *fused* into one indivisible being of *a man*, like a great living tree in which are bark, wood, and dry twigs; green leaves, flowers and fruit. Is it possible to conceive of and understand such a being by our reason?[9]

The attempt to visualize this marvelous conception of Ouspensky's will lead us into another space metaphor. We begin with a clever visual-conceptual device employed by Ouspensky and by many others, most notably Edwin A. Abbott in his book *Flatland*. This device has us try to picture higher-dimensional worlds by beginning with those of lower dimensions. To understand the fourth dimension and our difficulty in conceiving of it, Abbott writes about the imaginary adventures of a creature, a square, who lives in the two-dimensional plane of Flatland and who struggles to understand the world of three dimensions.

We must imagine this two-dimensional world as an exceedingly thin plane, for we cannot conceive of a physical existence of objects with no thickness at all. In this world, our perceptions are severely limited. We see objects only *on edge* as it were; we can sense only their periphery. For example, if we come across a circle, we can sense only its circumference and not its inside area. And even this circumference would appear as a line segment, as would all objects. We should be unable to perceive directly the curvature of the circle or the shape of any object in the plane. This can be pictured clearly by viewing from the edge of a flat table (or imagining doing so), a number of paper cutouts of squares, circles, etc., lying flat on the tabletop. It is all quite analogous to the limitations on our perceptions in the world of three dimensions. We are able to see only projected surface areas of objects, not their interior volume.

Now imagine the sheer two-dimensional plane of our existence being pierced and intersected by a normal tree from the world of three dimensions. You might think of the plane as the surface of a body of water covering about half of the tree. A good portion of the foliage rises above the water, so that our plane is cut through by a complex pattern of leaves, blossoms, fruit, twigs, and branches. All the parts of the tree that intersect our plane appear to us in cross section as plane figures of irregular shape whose peripheries, alone, we can sense. The entire cross-sectional array would seem like a collection of disconnected islands with no relationship among them. It would be almost impossible for us to realize that these clustered island-barriers are in reality all interconnected in the third dimension and, further, that they all participate in an incredibly complex system of mutual support and communication that characterizes a living organism. Even harder for us to

The Tree of Man

understand would be the relationship of each of these apparently isolated islands to the soil below, with its water and nutrients, all completely outside our ken. Or to the atmosphere and its weather, or to the distant sun.

By analogy, now, we may try to picture the collective humanity of which Ouspensky speaks as forming a single organism, a *tree of man*. We experience the various parts and members of this tree as if they were isolated individuals, each behaving like an independent agent with no apparent relationship or connection to the others. We see each person as an island in our three-dimensional world. We conceive of human relationship, interdependency, and community in terms of the somewhat abstracted and externalized notions of friendship, brotherhood, and love. These are things that somehow mysteriously bridge the gap of space between island-individuals, but not which link them organically. We can hardly begin to imagine that "all kings, all slaves, all saints, all sinners—all taken together (are) fused into one indivisible being of a man, like a great living tree." We cannot know of the fundamental unity of all people as a single complex organism in some higher-dimensional world. We do not recognize our mutual dependency and support, nor our relationship to the nourishing surroundings. Do our fingers recognize their mutual connection; do our blood cells know the vast organism they partake of; do we experience our participation in humanity?

All of this is by way of suggesting that our experience and conception of the human race is yet another spatial metaphor. The vast array of human relationships through blood, race, society, religion, and politics is normally conceived of in familial, psychological, social, cultural, and historical terms. But we may equally well imagine it as a spatial network—as the multiconnected cells, tissues, and organs of a great living organism. And Ouspensky's notion of our ordinary perception of the one "man" or "tree" in seemingly variegated, separated, and isolated parts is further clarified by the human spatial metaphor. For as we saw earlier, the very essence of space is to provide for separation, articulation, and distinction. The spatialization of man is tantamount to the peopleing of the world.

The Language of Space

As a final comment in this discussion of spatial metaphors, and the intuitive approach to space generally, I should like briefly to consider the unlikely matter of prepositions. That prepositions, as we use them in language, have a great deal to do with our intuitive notions about space is obvious as soon as we begin thinking about them. They are words used before nouns or adjectives, as *The Random House Dictionary* says, "that typically express a spatial, temporal, or other relationship, as *in, on, by, to, since*." Most prepositions have some spatial or temporal connotation: inside the box, inside an hour, under the rock, under a minute, into, from, above, below, within, and so on. Many prepositions can be used either spatially or nonspatially: by the brook, by William Shakespeare, for this reason, for a mile, for an hour. But it is certainly significant that the same word serves both purposes. And does there not often remain some trace of a spatial flavor, even in the alternate uses, for example, in "That book is *for* me." What relationship of the book to me does *for* express? Possession, nearness, acquisition? Is there not some implied motion toward me in the idea of "intended for?" Such spatial connotations may not be found in all prepositional uses, but they do occur in many of them. Many of our concepts concerning distance, orientation, relative position, and motion in space are communicated and felt through prepositions, which thus seem to incorporate the peculiarities and idiosyncrasies of our spatial metaphors.

Now the other thing that always comes to mind when I think about prepositions is the statement, which was drummed into my head many times over in the study of foreign languages, that the most idiomatic words in a foreign language, and therefore the most difficult to use, are the prepositions. Wilson Follett, in the article on prepositions in *Modern American Usage* is unmistakable:

> One of the greatest difficulties in learning the European languages is the mastering of the idiomatic use of prepositions with verbs, adjectives, and nouns. . . . A mastery of all the differences and subtleties is rare, yet nothing gives away the foreign speaker or the insensitive writer like

> the misused preposition, or again the tenable preposition which turns out to be wrong for the meaning intended.[10]

Why in French must we say *en France*, but *à Paris*? Or in German, *bei mir ist alles klar*, and yet *komm sitz bei mir*? These and countless other examples will occur to the language student. Just the idiomatic governance of the cases of nouns in Latin, German, and Russian is enough to make anyone's head spin.

Now what is peculiar and interesting in this is that the aspect of a language that is most idiomatic and characteristic, and, at the same time, most difficult for foreign speakers to master, is its rules for the use of words that describe spatial and temporal relationships. We learn our prepositions at our mothers' knees (not by them or before them), and heaven help any adults who must struggle with them later in life. But just as a language characterizes the state of consciousness of a people or a culture, as Barfield, Whorf, and others argue, so it is the indiosyncrasies of our spatial metaphors that prepositional usage reflects. The people of one culture and tradition experience and communicate about space differently from those of another. And since our space-constructs are the fundamental metaphors by which we represent our experiences to ourselves; since they are the *a priori* categories that we use to organize our otherwise chaotic perceptions; they must reflect our particular idiosyncratic and unique view of the world. It is this deeply felt view that our mother tongue evokes with its idiomatic use of prepositions. These common yet elusive little words conceal our most intimate beliefs about the nature of reality. Prepositions form the invisible linguistic scaffolding of space and time.

Chapter 4.

Time for a Change

In this chapter we shall be concerned with time—with its subjective and intuitive character. The inner experience of time contrasts markedly with that of space, and yet we often spatialize time. The spatial approach is frequently used in the more objective handling of time, in particular in its measurement. When we speak, for example, of time intervals and durations or of time order and sequence, we have in mind an imaginary long straight axis of time with points on it locating events and distances along it measuring the elapsed time between events. The very words *interval, duration, sequence* evoke spatial images that help us think about time and its measurement. In other words, for quantitative and related conceptual purposes, we picture time as a kind of one-dimensional continuous space. It is in this spirit that the theory of relativity treats time as the fourth dimension, added to our physical three-dimensional space. And one finds this spatial view of time throuugh-out scientific literature.

Since we are primarily interested in nonquantitative aspects of time, it might be thought that the spatial representation of time is not relevant. But this is not so for two reasons: First, as we have seen, there is a subjective foundation to measurement and quantification. This means that even the spatialization of time has a subjective basis. Second, the boundary between an external spatial time and an internal intuitive time is blurred and often nonexistent, as in the case of the puse or heartbeat.

We begin with a spatial metaphor for time which is revealing itself and is also the foundation of some examples to come. Once again we borrow the very picturesque imagery of Ouspensky.

The Time Tunnel

Imagine that you are a wormlike creature with a human mind. You are wriggling through a tunnel in the earth. You have no sight or hearing, and, to simplify matters, let's assume you have only one sense, that of touch which is located at the very tip of your front end, your nose, as it were. Your sensations consist entirely of what you feel with your nose as you come upon things in your travels through the tunnel. These sensations are experienced one at a time and in sequence, so that your overall conscious impression is a kind of *now this, now this, now this, and so forth*. You perceive the world as a train of events in time. By a heroic feat of intellect, you might learn to conceptualize these events as a series of points along a one-dimensional continuum (the tunnel), by becoming aware of your own motion along that continuum and taking it into account. But failing this, you would have a sense of time only, and not of space. Your impression would be entirely of a sequence of unanticipated events which seem to come to you in time. In effect, as far as your sensations are concerned, you would live in a world of zero dimensions, since that spot of awareness at the tip of your nose is like a geometrical point with no spatial extension. (In geometry a volume is three-dimensional; a surface, two-dimensional; a curve or line, one-dimensional; and a point, zero-dimensional.) We might say that time is your dimly perceived experience of the one-dimensional space you exist in but do not apprehend.

Now imagine adding sight and body sensations to your sense repertoire, and that the tunnel is lighted so that you can see. You are aware of things ahead of you in the tunnel and of your body's length along it. You have a direct sense of the one-dimensional space you inhabit, but with a rather limited perspective. Let's idealize the situation so that you are a vanishingly thin creature with a length but no thickness or width. And we imagine the same is true of the tunnel. You are an exceedingly fine filament-creature moving through an equally fine bore. All you can ever see is a single point, the boundary or end of whatever is next in line ahead of you along the tunnel. You cannot sense distance by perspective since a point has no size and does not appear diminished at large distances. You can estimate distance only roughly in terms of the travel time taken to reach the nearest object, with your nose telling you when you arrive. As a member of a colony of worms inhabiting the tunnel, you have contact only with your nearest neighbor in front of you and behind you.

Now suppose that some foreign creature enters the tunnel just in front of you by breaking through the wall and that you can sense it as something new by virtue of its color. For instance, the eternally blue point in front of you suddenly turns red. What a shock! In your world, juxtaposition is destiny. Your space does not allow for objects to pass by one another; there is only one dimension. You cannot conceive of anything coming laterally from outside the tunnel. The creature completes its passage across the tunnel, and the red spot disappears. More incomprehension! Was it just your imagination? How is it to be explained? All you saw was the ever-blue point before your eye turn red and then blue again. It was for you an occurrence *in time*, a red interruption in the enduring blueness. Some strange thing appeared, existed briefly, and then disappeared. You experienced as time what in fact was the *length* of something crossing through your space. But you cannot comprehend this. As with the more limited zero-dimensional worm, time is your dim awareness of an unperceived space of higher dimensions.

The analogy carries over quite easily into higher-dimensional worlds. Imagine now a cigar-shaped object passing through the two-dimensional plane of Flatland. Only the circular cross section of the cigar in the plane is perceived, seen edge-on as a line segment by the Flatlanders, of course. The passage of the object through the plane seems like a strange sequence of events in time: the appearance of a point out of nowhere, the blossoming of the point into a circle which swells to a maximum size, decreases, shrivels up to a point once again, and then disappears. It appears to be the life history of a circular creature which is born, matures,

The passage of the object through the plane seems like a strange sequence of events in time.

ages, and dies. The Flatlanders cannot understand that their experience of each phase in this life story is just the geometrical cross section of an object that neither is born nor dies, but simply *exists* in a space of higher dimensions.

By the same token, we three-dimensional creatures can view our time sense as an imperfect perception of the fourth dimension. All growth, decay, evolution, development, creation, destruction may be thought of as our apprehension of the sequence of three-dimensional cross sections, of four-dimensional objects passing through our world. It is as if the light of each individual human consciousness passes along its four-dimensional body, illuminating only a three-dimensional slice as it goes. And this is how we experience our own mortal lives.

For Ouspensky, these provocative kinds of images and dimensional analogies are really a prelude to his model of enlightenment, which involves the expansion of human consciousness to full four-dimensional awareness. He conjectures that in parallel with expanded space perception is a radical opening up of the mind to new vistas of understanding, knowledge, and revelation. The resulting state of mind, according to Ouspensky, is tantamount to that of the great seers, mystics, and prophets of history who have experienced transcendent enlightenment.[1]

I am in great sympathy with Ouspensky's approach, although I believe that his general view of space and time is basically idolatrous. Why, for instance, should the fourth dimension be an ultimate? Would not an intiate into the practice of mind expansion soon seek the fifth dimension, and then the sixth, and on and on ad infinitum? Then again, is there not an irreducible aspect of time at each step of this dimension hopping that is not explained in terms of the next dimension? The worm, in his tunnel, already has a time sense connected with his own motion, apart from the mysterious intrusion of any higher-dimensional creatures. Besides, is space to be conceived of as a final explanation for anything, rather than just as one metaphor among others? But I should not be too hard on Ouspensky. In his notion of ever higher dimensions of space may be buried the view of ultimate transcendence of space and time altogether. Ouspensky, himself, seemed to repudiate *Tertium Organum* later in life. I should not go that far, for the geometrical creations of that book are marvelous space-time metaphors.

The Ouspenskian space analogies of time are particularly helpful in bringing out our spatialization of time in such notions as duration, interval, and sequence. Before we saw that the spatial view of time was connected with quantification and time measurement. We shall now see its link with our more intuitive sensation of time. Imagining time as an extra dimension of space, of which we are but feebly aware, gives us a glimpse of how we may actually experience time. But it also shows us our usual linear, geometric time notion as an abstraction and simplification of our direct, subjective *qualified*[2] experience of time.

Creation and destruction, growth and decay are basic to our sense of time. It is the blue dot turning red, the circle being born and dying, the history of a human life that is our primary experience, not some linear, relentless flow along an abstract, external, vacant axis. Our time sense has quality, texture, character. It is uneven, cumulative, cyclic. It has meaning, value, purpose, life. From all these inner, organic, human experiences we have extruded an external time, like a smooth, characterless wire stretched taut from infinity to infinity. We must seek out the myths and metaphors that will give time back its sinew and life, that will transform the unadorned wire back into the bejeweled necklace.

Marking Time

Although we conceive of time as linear and smooth flowing, our ability to measure temporal duration depends on our recognition of a periodic time structure in natural phenomena. The repetitiveness of the earth's rotation on its axis and passage around the sun, or of atomic vibrations, provides a standard time interval, basic to all measurement. But which is the primary experience—smooth time or its punctuating rhythms? Our usual objective view of time puts the cart before the horse. We are not presented intially with some pristine, abstract, unqualified kind of time, served up, as it were, on a silver platter like some bland, tasteless dish which we then embellish with spices and herbs to give it the flavor and character of periodicity and repetitiveness, so much tastier to our temporal palates. We do not qualify an originally vacant time. On the contrary, we abstract an empty, linear time from a qualified, organic one. We begin with the experience of cycles and periods, of change and variation, of growth and development. And it is this more fundamental and subjective sense of time which is embodied in our legends, myths, and esoteric philosophies.

Astrology again provides a clue. Previously we saw that astrology characterized a state of consciousness in which no sharp distinction is made between inner and outer space, in which no clear notion of space as we conceive of it may even exist. We can also recognize in astrology a kind of temporal consciousness very different from our own. Each moment in astrological time is stamped with a unique configuration of the heavenly bodies, a moment which contains like a seed the potential for future development. The newborn infant with its myriad human traits and the natal star pattern at its birth are two different ways of characterizing a certain moment in time. They provide a symbolic signature of that instant. To astrological consciousness, the passing moments have quality, character, import, potential. But, to talk of a moment in time is already to begin the abstraction process. The direct temporal experience is simply of the childbirth and pattern of the heavens.

The *potential* nature within the child-star configuration is manifested to us in time as a process of unfolding and developing, a sequence of stages, a history. The potential character inherent in the moment becomes actualized in time. Perhaps in some higher state of human consciousness, one would not experience this apparent evolution; one would see the totality, the whole cigar, as it were, and not only its changing cross sections. The separate symbolic moments of astrology would seem simultaneous to us. Even without such revelations, we sense the basic, organic, qualified nature of our experiences as fundamentally different from a series of empty, identical points or instants of time along an axis. Astrology helps us see the passing moments not as vacuous, but as full of meaning, quality, and value.

When we look at periodic time experiences, we see, too, that they do not have the simple, precisely duplicated character that the objective, linear view of time leads us to believe. Planets never perfectly retrace their orbits nor traverse them in identical time spans. They disturb each other, they wobble and slow down, and their orbits are perturbed and precess or rotate in space. All these effects are very slight and barely observable, but have nonetheless caused problems, for example, in astronomically defining the second. Far more significant, however, is the fact that the orientation and time periods of the orbits bear no simple relationship to one another. The planets are not in step with each other. Thus the overall configuration of the planets never repeats itself. Some rare coincidences, like the lining up of four or five planets, may have been repeated in recorded history, but each instance will make its own unique imprint. For instance, if Mercury, Venus, Mars, and Jupiter become aligned every few centuries or so, they will always be found in different regions of space and will be accompanied by different configurations of the other planets. This is reflected astrologically in the human characteristics attributed to these repetitive, but never identical, star patterns. All individuals have much in common, yet each is unique.

This is how it is with all periodic human experience. The seasons of the year recur endlessly, yet no two springs are alike. With each passing year, we are older, different, but somehow the same. We eat thousands of dinners in our lives, all different and all similar. It is repetition with variation that delights and fascinates us. And we have celebrated this fact throughout history. In our holidays we rekindle the memory of some significant event and strive to relive or repeat it through festivity, ritual, commemoration, and reflection. Our clocks, calendars, and schedules reflect the endless repetition of the hours, days, weeks, and years, whose passing will seem hollow and relentless to us only when we fail to fill them with human meaning, value, and quality. Again, I state the case in reverse. For it is the quality of experience that precedes the continuum in which we imagine it to exist. Time is no empty container we fill. Linear, external time is abstracted from the organic, human temporal experience; the former is a simplified metaphor for the latter.

Incidentally, while on the subject of holidays, I might point out that there exists an *inner* connection between number and time. We treat certain commemorative anniversaries as somehow more special or significant than others, for example, a 25th wedding anniversary as compared with a 24th or 26th. This symbolic relation of time and number is also reflected in allusive numerological systems, like playing cards and the *I Ching*. We shall return to this matter in more detail when we come to number metaphors.

Theme and Variation

There is no metaphor that better illustrates and celebrates our insatiable and joyful appreciation of repetition with variation than does music. Music has a deep connection with time. Rhythm, tempo, and syncopation, for instance, are temporal. Even pitch is basically temporal since it depends on the frequency of a sound tone, i.e., the number of oscillations per second of the air column or the string or the reed in a musical instrument. Thus our musical sense of interval, melody, key, and harmony depends on time. In fact, time is usually thought of as the *medium* of music in analogy with the role played by a canvas or two-dimensional surface in painting and by three-dimensional space in sculpture. The notion of a medium again suggests some empty continuum (time or space) which exists prior to the act of creation and which the artist subsequently *fills* with his or her work. Is the painter filling space or creating it? By now it should be clear that I favor the latter viewpoint—that space is a creation of consciousness, or more properly, an inseparable part of it. I want now to broaden this idea to encompass music and time.

Basically, music is a metaphor for qualified experience. The nature of music is not that of uniformity or constancy but of pattern and ordered variation. An indefinitely prolonged note or period of silence has no musical value. (Even the "silent" works of John Cage are not pure vacuous experiences but involve some anticipatory or implied interaction in time between an audience and performing artist.) Music implies sound variation with a pleasing or novel effect, although what is considered to be pleasant and new musically varies among cultures and historical periods. One of the most obvious musical patterns is melody, a sequence of notes in time that is often described as a *horizontal* pattern. Harmony and counterpoint, on the other hand, are part of the *vertical* rather than the horizontal structure of music. Vertical and horizontal refer to the representation of musical composition in its written form on a page. The melodic unfolding of a piece of music in time is represented by the horizontal sequence of notes on the musical staff, whereas the simultaneous notes and chords of harmony are represented vertically. Thus all musical form is either horizontal-melodic or vertical-harmonic.

These two dimensions of music reflect the two fundamental characteristics of our experience of time: sequence (unfolding) and simultaneity (parallelness). If time is thought of as a kind of rope, melody is its extension and length, while harmony is its combined strands and strength. But a rope is too simple and uniform to serve as a proper metaphor for time or music with their varying, ever-novel qualities.

Harmony, polyphony, counterpoint, and accompaniment all refer to the simultaneous sounding of different tones which blend or contrast with each other into a pleasing synthesis of sound. The very word *harmony* suggests an aesthetically satisfying combination of things in an ordered pattern. When things harmonize with each other, they not only go well together, but also enhance, nourish, and elevate each other so as to produce a whole greater than the sum of its parts. A harmonious pattern is an integration, a synthesis, a unity. It qualifies experience with texture and character; it imbues it with meaning and value. We cherish the harmonious experiences in which what we experience and what we perceive seem to combine into a soul-satisfying unity. In music, the artist consciously creates such a harmonious unity. The vertical structure of music and the simultaneous texture of human experiences are different metaphors for the same thing, only part of which is abstracted into our objective notion of time.

The more complete and subjective time experience is an organic and varying complex of things. Human experience may not always be harmonious, but it is nevertheless compound. Many things go on together which, according to circumstances, may disturb or please us. Even the quest for the "simple" life is more a search for a harmonious combination of experiences than for fewer of them. Communing with nature is like apprehending a jewel of many facets. Simplicity implies singleness in the sense that a harmonious combination is a synthesis or a unity. To experience the many in the one is an exalted state of life, and music too strives for this quality of exaltation. A harmonious pattern is a metaphor for the experience in which all things blend perfectly into one. It enables us to sense and appreciate some ultimate state of order and perfection. Harmony in music allows us to partake symbolically of that state and to sense it even in a limited and finite form. It soothes our fears of chaotic disorder. It gives us a glimpse, even in the fleeting instant, of the ever-present beauty, meaning, and order of our complex world.

Horizontal melody is an equally profound metaphor for human experience. We really cannot separate the horizontal from the vertical in a harmonized piece of music. Music is experienced as a unity. It may be broken down in analysis, but never in performance or appreciation. One cannot even imagine the harmony of sound without some horizontal extension in time. Sound does not exist in the instant, without persistence and duration. Here we see the deep analogy between the music and the time sense. For music, like human experience, cannot exist at all in a pointlike, instantaneous present, but must partake of both the past and future for its being. A pure tone cannot be frozen in time; there is no sound without motion, without vibration. The very quality or pitch of a note can neither exist nor be heard unless it endures for some finite time interval, however small.

The most basic melodic form is theme and variation, the statement of a simple melody followed by a series of recognizable or related modifications of the original theme. All melody patterns, from the simplest A-B-A songs of popular music to the most complex sonata forms of classical and romantic music, are derived from the model of theme and variation. What is so meaningful and delightful about this melodic form that composers should never tire of using it, nor listeners of hearing it? What archetypal human experience does it symbolize for us? To begin with, theme and variation is a horizontal musical structure, which means it is a metaphor for the flowing or progressive aspect of time. We have a basic human sense of the passage of time, and our experience tells us that this passage is characterized by change and metamorphosis which on occasion may be unanticipated and frightening. The ultimate prototype event of an unexpected and unwelcome nature is death. But the many experiences of creation, destruction, and chance occurrence may be either a delight or a threat to us. The rope of time seems to be woven from two kinds of strands: the smooth silk of causal order and the rough hemp of chaotic chance. Our passage in time is never secure, and we constantly seek the safety of causal events to reassure ourselves that life may be comprehended and controlled.

On the other hand, it is not dull uniformity we seek, either. An experience of perfect unvarying constancy not only would be boring and monolithic, but might not even constitute existence as we know it. Rather, what we seek and enjoy most is the novelty of variation within prescribed limits. The withering and dying processes of fall and winter are acceptable, even beautiful, to us, for we are familiar with the *ordered* steps of seasonal metamorphosis. A game of chess is a fascinating challenge because of the endless possibilities within its ordered rules. Travel is exciting because it exposes us to people, places, and cultures that are new and yet bear an understandable relationship to their familiar counterparts at home. The ideal is not simply a repetitive or periodic pattern which may be initially interesting, but eventually turns dull. Perfectly symmetric designs are ornamental but not usually moving. It is variation with regularity that provides the necessary novelty and interest to assure us of excitement without catastrophe. In music, it is theme and variation that offers us an exerpience of meaningful novelty within a comprehensible, ordered structure. Theme and variation is a basic metaphor for a life of secure delight and growth, of ordered, novel change. The recurrence of melody in an ever-refreshed and -delightful form is like the cycle of the seasons: ever old, ever new.

What a marvelous moment it is in the sonata movement of a Beethoven symphony when we reach the recapitulation and hear the familiar theme, made so much more meaningful than at its initial appearance by Beethoven's profound elaboration of its thematic and rhythmic content in the development section of the movement. Or after Bach's thirty exquisite *Goldberg Variations*, when we encounter once again that deceptively simple and stately theme from Anna Magdalena's Notebook, now brought to life by Bach's probing, moving inventions. Indeed, hearing the profound variations and development of a theme by a Beethoven or a Bach is like witnessing the unexpected but understandable blossoming of a magnificent flower from a simple seed whose fantastic potential we could hardly have ever imagined to exist. The elaboration and unfolding of a theme in music follows from what is implicit or potential in the original melody, just as the stages of growth of a living organism evolve from a seed with its hidden potential. (And if the development of a musical theme requires the genius of a creator, shall we say less of the growth of an animal or plant?) The meaning of music cannot be fathomed or derived from something else, for music is a metaphor, on the same footing as our constructs of reality. All the vertical and horizontal complexity and texture of our time metaphor is reflected in music.

We may probe the metaphors even further to reveal the abstract, simplified nature of our objective time construct. A theme in music, an astrological natal chart, and a seed in the organic world all have in common the notion of a potential which can unfold or evolve in time. But why is the *passage of time* evoked to explain these things? Is it any less of a mystery that an acorn should *turn into* an oak *in the course of time* than that the acorn should already contain the oak *in potentia*? How can our vacuous, linear, one-dimensional notion of time possibly account for the fantastic metamorphoses and changes we witness around us? Of course, we don't claim that time is responsible for these changes, but we do picture time as a kind of medium in which changes transpire. And physical science assumes as a matter of course the idea of the continuous evolving of stages in any natural phenomenon during some interval of time. Why does getting from one end of a time interval to another make it seem reasonable that one thing has disappeared and another has taken its place? Time may have been a most fortuitous purchase for the expression of physical law, but into the bargain with it we have had to accept mortality and death. Seed and flower, child and adult, theme and variation are not really a matter of creation and destruction, but of unrevealed potential. In chaotic unity, all things exist and all things are possible. Time, like space, is a way of spreading out and articulating all these potentialities and actualities. Music, like astrology, reveals the deeper nature of time and space.

Eternity in an Hour

Our next time metaphor deals with compacted and intensified time. First, contemplate a piece of music *in toto*, not as experienced in performance, but as held in the mind of a listener or musician thoroughly familiar with the piece. In what state is such a mental musical work in the memory? It is not in time as in a performance, any more than our ideas, recollections, and dreams may be thought of as existing in objective time. We may, of course, experience portions of the piece in subjective time when we mentally run through or hum certain passages. But I'm talking about how we apprehend and appreciate it as a whole, for instance, after having heard a satisfying performance of it or having read through its score. (Incidentally, in what state is music as a written score? Spatial?) With a really meaningful and beautiful musical work, one has a sense of it as a unified thing with a character and value of its own, like a cherished friend. One could not begin to say all that it means, or to define it. It is held in the mind as a unique, known, and valued entity. The full work is felt, not as evolved and laid out in time, but as capsulized and unified. It is as if, in our Flatland illustration, we see the whole cigar in three-dimensional space and not just its evolving cross section as it pierces our two-dimensional plane. The same is true of a movie or novel after seeing it or reading it. We may contemplate the incidents, mentally converse with the characters, imagine alternative twists of plot; it is a living organism for us with which we can communicate, a total experience, not something strung out in a sequence. But to me, the musical metaphor is more profound, for we cannot translate musical meaning into other terms as easily as we can with literature. Whereas literature is a more direct outgrowth of life, music stems from the common root of human meaning at a deeper, more inchoate level. The apprehension of a piece of music, held as a totality in the memory, gives us a glimpse of a different kind of time consciousness from ours, one in which experiences are not sequential and stretched out in a line, but are simultaneous, and amalgamated into an organic complex.

This cumulative notion of time is beautifully illustrated by Benjamin Lee Whorf in his book *Language, Thought, and Reality.*

In the essay entitled "The Relation of Habitual Thought and Behavior to Language," Whorf discusses the Hopi practices of covert participation and mental collaboration. To ward off evil and dispel the tendency toward social disintegration, the Hopis employ a practice of "preparing" which involves the "great power of the combined, intensified, and harmonized thought of the whole community," leading to a "rather remarkable degree of cooperation that, in spite of much private bickering, the Hopi village displays in all the important cultural activities." Whorf continues:

> Hopi "preparing" activities again show a result of their linguistic thought background in an emphasis on persistence and constant insistent repetition. A sense of the cumulative value of innumerable small momenta is dulled by an objectified, spatialized view of time like ours, enhanced by a way of thinking close to the subjective awareness of duration, of the ceaseless "latering" of events. To us, for whom time is a motion on a space, unvarying repetition seems to scatter its force along a row of units of that space, and be wasted. To the Hopi, for whom time is not a motion but a "getting later" of everything that has ever been done, unvarying repetition is not wasted but accumulated. It is storing up an invisible change that holds over into later events.[3]

Whorf further elaborates these ideas in a footnote, which is particularly relevant to the themes of this book:

> This notion of storing up power, which seems implied by much Hopi behavior, has an analog in physics: acceleration. It might be said that the linguistic background of Hopi thought equips it to recognize naturally that force manifests not as motion or velocity, but as cumulation or acceleration. Our linguistic background tends to hinder in us this same recognition, for having legitimately conceived force to be that which produces change, we then think of change by our linguistic metaphorical analog, motion, instead of by a pure motionless changingness concept, i.e., accumulation or acceleration. Hence it comes to our naive feeling as a shock to find from physical experiments that it is not

> possible to define force by motion, that motion and speed, as also "being at rest," are wholly relative, and that force can be measured only by acceleration.[4]

Indeed, if we had had the Hopi intuition of time in which acceleration is recognized as qualitatively different from uniform motion, the whole Newtonian revolution might not have been necessary. Newton and Galileo carefully distinguished between uniform motion (at constant speed in a straight line) and accelerated motion (at varying speed and/or in a changing direction). And it is this distinction along with the spatialization of time that lies at the heart of classical physics and of our conventional modern concept of time and motion. (Note Whorf's point that acceleration need not be thought of in motional terms at all.)

All this has now been changed by Albert Einstein (at least the physical theory has, if not our concepts). In general relativity theory, the need to distinguish between accelerated and uniform motion is eliminated. All motion is seen as uniform in Einstein's curved spacetime. The seemingly irregular motion of objects is explained, not in terms of forces acting across space, but simply by the geometrical properties of the non-Euclidean spacetime we live in. Not only is all motion relative, it is uniform; it is space that is curved and nonuniform. Modern thought severs the notion of acceleration from its temporal matrix and grafts it to the Hopi metaphor of accumulation and growth.

The experience of repetition leading to accumulation, growth, fruition, rather than our more usual feeling of a meaningless unvarying recurrence, is not at all foreign to us as an alternative metaphor to linear sequential time. This alternative is illustrated by astrology, music, ritual celebration, cyclic or seasonal experience, and Ouspensky's spatial metaphors. Many other examples occur in cultures and times different from our own. One that is particularly difficult for the modern Westerner to understand and accept is the concept of reincarnation, a belief that pervades Hindu philosophy but which can also be found in early Greek thought and in many so-called primitive cultures. Trying to conceive of reincarnation in linear time presents problems for us. Yet if we were able to think with an organic cumulative view of time, we might see reincarnation differently, as the soul's growth toward perfection, or at a deeper level, as the total potentiality of each soul. Rather than suggesting an iterated series of lives stretched throughout history, the idea of reincarnation may really hint at something more like a perfect jewel, each of whose many facets can reflect only a part of its total harmonious brilliance.

A final illustration of the metaphor of fruition-through-repetition is one that we can immediately appreciate—the sex act. Surely, the repetitive steps we take to achieve orgasm do not strike us as boring!

Fascinatin' Rhythm

Perhaps the most obvious temporal characteristics of music are rhythm and tempo. In the light of the last section, it might seem that rhythm and tempo represent a repetitive sequential element in music, akin to linear time. They seem to form the substrate of music in the same way that we picture time as the medium of experience. But this conception is part of our usual temporalized view of music and spatialized view of time. Is rhythm really the analogue of an even time flow, or does it already embody a structure from which we abstract linear time? I believe that an exploration of the inner experience of rhythm discloses another fundamental metaphor for time. At the core of rhythm we find a polarity, characteristic of the first mysterious creative step in passing from unity to duality, from the potential to the manifest.

Any attempt to treat this truly unfathomable mystery, the connection between the primal act of creation and rhythm, is extremely difficult. Fortunately, our understanding is enlightened by Sri Krishna Prem in a very penetrating discussion of the Hindu *Creation Hymn* of the *Rig-Veda*. This poem and sacred text stem from a civilization in India, more than four thousand years ago, characterized by human consciousness very different from that of today's Western world. The first two stanzas of the poem follow:

> There was not non-existent nor existent;
> there was no realm of air, no sky beyond it.
> What covered in, and where? and what gave shelter?
> Was water there, unfathomed depth of water?

> Death was not there, nor was there aught immortal:
> No sign was there, the day's and night's divider.
> That One Thing, breathless, breathed by Its own nature:
> Apart from It was nothing whatsoever.[5]

In the course of his discussion, Prem comes to the line "That One Thing, breathless, breathed by Its own nature," and in part makes the following comments on it:

> . . . in the One . . . there is a certain rhythmic process symbolized as breathing. Within the One exist in potentiality the two poles of Subject and Object. . . . We may conceive that "breath" as a rhythmic alteration of polarity between these two. When the pendulum swing has attained a certain amplitude we may conceive something happening analogous to what takes place in ourselves when mental processes that have been going on below the threshold of consciousness rise into the light and emerges as "I am experiencing such-and-such."[6]

Prem suggests rhythm as anterior to or possibly synonymous with duality. (We mustn't get hung up on time priorities, for as Prem says, "It must be remembered that we are in a region beyond time as we know it and the stages are more logical than temporal ones."[7]) We may conceive of a subtle, subliminal, fluttering process as the initial creative act of breathing, rhythm, differentiation. It is reminiscent of an embryo gradually developing its inner beats and pulsings only to burst into the first breath of life. I cannot help but think, too, at the opposite end of life's time scale, of the last fluttering, bubbling breath of a dying person.

Rhythm is a differentiation—an articulation of the unified chaotic primal experience. It is often represented in modern mathematics as a wave form in time. Viewed as such, it still appears as a continuous line but is no longer straight. It curves up and down in the typical crests and troughs of a wave. Our sense of rhythm, then, is a *singling out* and accenting of these high and low points on the wave. We qualify these extremes, give them significance, and perceive them as the beats of a rhythmic pattern. The stages, if we may think in such terms, in the process of differentiation and the evolution of rhythm are the following:

1. An original state of undifferentiated chaos.

2. The original state either transforms itself into or is newly perceived (which implies a subject/object dichotomy) as one having extension and duration.
3. The sheer uniform continuity is qualified and takes on a variable periodic structure (a wave).
4. The extremes of the wave pattern are isolated, articulated, and separated as distinct entities with individual identity (beats in a rhythmic pattern).

Rhythm appears as a discrete organizing pattern. Duality and rhythm are somehow associated with discerning a quality, a variation, a nonuniformity within a flat uniform experience.

It is *we* who single out the extremes of the motion of a pendulum as special, isolated, discrete. But to some imaginary mind in the pendulum bob, its motion would seem one-dimensional, continuous. Or a worm with only a sensory nose moving along a wavy tunnel would sense nothing special about the high and low points. It cannot distinguish between straight-line and wavy motion. In fact, it cannot even detect its own motion. This worm's consciousness might be categorized as somewhere between stages 1 and 2 above.

Rhythm and music are qualifications of experience which we apprehend. And quality, differentiation, articulation are the basic characteristics of a dualistic or multiple existence. The meaning of music lies deep indeed. We recognize in it the primary creative act, the act of breathing, inspiring, pulsing, distinguishing, producing two and many from the One. At an even deeper level is the continuity and duration of time, itself, basic to both the straight line and the wave. The passage from step 1 to step 2 bespeaks this mystery. How do we go from nothing to something, from the collapsed point of zero-size to the stretched-out line of finite extension or duration? Once we have a one-dimensional line, we can create quality, curvature, variation, and then evoke duality and multiplicity. But first the point (chaotic, unified, dimensionless, timeless experience) must become the interval. How? This is the fundamental act of creation, of the movement from unmanifest to manifest. Buried deep within music and rhythm is the secret.

Entropy and Eschatology

Having discussed the relation of music, rhythm, and time to the creation of the universe, we shall take a brief glance now at the other end of the time scale, where we find destruction and doom. As much as physical scientists might prefer to avoid such speculative matters, they are saddled with a concept that does imply a certain final state of things. This is *entropy*, which is closely associated with our intuitive idea of the flow of time. Entropy is a measure of the amount of disorder in a system, which, according to the Second Law of Thermodynamics, always tends to increase with time.

Physicists mean by the *order* of a system the structuring in space and time of the elements of the system. For example, a postal employee is creating a spatial order in a system of letters when pigeonholing them according to zip code. A painter creates order on a two-dimensional surface with pigments. And a telegrapher produces an ordered pattern of signals in time when sending a message. What the Second Law tells us is that the natural tendency of things is to progress toward states of greater and greater disorder. It is more natural for all the books in the library to be heaped in a great pile than for them to be neatly arranged on shelves according to the Dewey Decimal System. Or, to give an example from physics, it is more likely to find a gas, confined to a box, to be homogeneously distributed throughout the box and to have a uniform temperature than it is to find some of the gas bunched up in one corner of the box and very much hotter than the rest of the gas.

Entropy and the tendency for increasing disorder are related to probability, for what the Second Law really says is that the *likelihood* of a disordered state is greater than that of an ordered one. And the natural spontaneous processes in nature are thus understood to produce the most probable states of matter systems. The most probable end state of a burning stick consists of smoke, hot discharged gases, and charred wood. It could conceivably become a carved doll, but this is extremely unlikely.

Time gets into the act because the law of entropy increase implies a one-directional flow of events. Since the original stick is in a more ordered state than its burned remains, the process of burning cannot be reversed. That would imply an *increase* of order, a clear violation of the Second Law of Thermodynamics. Now if we carry all of this to its logical conclusion, we project a picture of the universe in which all things are supposed to reach their highest possible states of disorder. Applying this to the universe as a whole, everything in creation would presumably reach some final maximum state of disorder. (This idea was popular in the late nineteenth century and influenced writers and artists.) And since order is characterized by a structuring of space and time, the final, maximum disordered state of the universe would lack all structure. There would be no distinction, extension, or identity of any kind. In short, there would be no space and time, and we should find ourselves in a zero-dimensional chaotic point. The flow of time inevitably leads to chaos.

Now the most prevalent cosmological theory today, the big bang theory, presupposes that the universe as we know it has evolved from the explosion of some primal state of matter and energy, called the fireball, at the beginning of time. This is an inconceivably concentrated ball of matter and energy at an unimaginably high temperature and pressure. Cosmologists do not describe its composition very well before it had expanded and aged by some minutes, at which time it is assumed to have consisted only of neutrons, protons, and radiation. One can make conjectures about its nature at earlier moments, but at time zero it is simply indescribable. It is not even clear whether the primal fireball existed *in* time and space or *contained* time and space, for there is no possible operational meaning to the idea of space and time beyond the material limits of the universe. In other words, this fireball sounds suspiciously like both the chaotic point at the end of time and before time in the Hindu creation myth (although we must admit to a certain cosmic *chutzpah* in venturing to identify it.) Time is revealed again as a polarizing separating metaphor, which divides the unified cosmos into initial creation and final destruction phases. Time appears as a kind of gap in the primal chaotic experience which somehow gives us the opportunity to apprehend and appreciate the diversity within the unity. Is this our role, the Grand Appreciator, with time as the mediator of our apprehension?

Just Because

As a final example of a time metaphor, let us consider the concept of cause and effect as treated in physical science, contrasting it with the synchronicity of acausal events. In the process, we shall become involved in a little relativity theory.

We have already had occasion to mention *synchronicity*, the principle of acausal relationships or connections between events and occurrences, which Carl Jung used to rationalize the *I Ching*. As a general principle, it also characterizes astrology, ESP phenomena, clairvoyance, déjà vu, and symbolic coincidences. In other words, synchronicity describes the relationship between two events that have a common or similar symbolic or semantic content but which do not necessarily have any causal connection.

Now I want to explore in some depth the metaphor of synchronicity and its relationship to time.

The only kind of connection between events that is of interest in physical science is a causal one. Every effect is assumed to have its cause. Wood burns in the presence of oxygen when sufficient heat is applied. Heat and oxygen are the combined cause of the effect called burning. Planets move in ellipses because the sun exerts an attractive force on them. Gravity is the cause of the planetary trajectories. One can go on indefinitely, for causality is the lifeblood of the sciences. Every physical law provides some rational connection between certain causes and their effects. When scientists are puzzled about some phenomenon, it is usually because they don't know its cause, and they may spend many years searching for this cause since they firmly believe that one must exist. Causality is the antithesis of randomness, chance, spontaneity. It is the element of order in time which is so essential to our sense of harmony and rightness about the universe. Causal time, like structured space, provides a metaphor of order that wards off the fear of chaos and death. Earlier in the book, we saw how science is constructed so as to form a self-consistent and predictable system. Causality is an essential part of that construction. Tobias Dantzig,[8] the mathematician-philosopher, bluntly states that our fear of the spontaneous biases our scientific theories toward evolution. According to Dantzig, even in cosmology and sociology, we recoil from the cataclysmic and we reject the possibility that revolution, spontaneity, and accident are dominant factors in the pattern of life and the cosmos. Causality, together with evolution, shapes our past, present, and future into a smooth, unbroken stream and protects us from the horror of the unexpected and the chaotic.

The affinity in our minds between these hazy notions of causality and continuity is so great that we habitually summon one to justify the other. But this is inevitable, Dantzig says, when we realize that our dogmatic trust in continuity and causality is basic to our intuitive sense of time.

It seems remarkable that two of the cornerstones of modern science, the theory of evolution and quantum theory, have managed the incredible feat of basing a system of order and reason on a notion of randomness. The marvelous hierarchical order of life on earth with its sequential development, and the highly symmetric and regular structure and behavior of matter, are ultimately traced to mere chance. The spontaneous mutation guided by natural selection, of living things and the random activity of atomic systems are taken as the alien raw material out of which ordered life and matter have been forged. How can chance be the basis of order? How indeed? Owen Barfield, in his criticism of Darwin's theory, writes eloquently about this ironic contradiction in modern thought:

> It was found that the appearances on earth (fossils, e.g.) so much lack the regularity of the appearances in the sky that no systematic hypothesis will fit them. But astronomy and physics had taught men that the business of science is to find hypotheses to save the appearances. By a hypothesis, then, these earthly appearances must be saved; and saved they were by the hypothesis of—chance variation. Now the concept of chance is precisely what a hypothesis is devised to save us from. Chance, in fact, = no hypothesis. Yet so hypnotic, at this time in history, was the influence of the (scientific) idols and of the special mode of thought which had begotten them, that only a few—and their voices soon died away—were troubled by the fact that the impressive vocabulary of technological investigation was actually being used to denote its breakdown; as though, because it is something we can do with ourselves in the water, drowning should be included as one of the different ways of swimming.[9]

There is something very ironic in all of this, for if causality is the temporal ordering mechanism that secures us against chaos, it has also given to time its quality of relentlessness and inevitability so that we feel ourselves being dragged kicking and screaming toward inescapable death. Our linear, sequential, causal time is like a narrow one-way road to doom. For the primitive mind with its cyclic, spontaneous, cumulative sense of time, the voyage of life seemed more multidimensioned, more like a ship cruising the seas than a train on a one-dimensional track. Life had its storms and occasional shipwrecks, but it also had its adventurous changes of course and familiar ports of call. Have our modern space, time, and causality brought us any real emotional and spiritual security? Have they helped to reveal the mysteries of our existence? What purpose have they served? Why have we invented them?

Oddly enough, modern science in the special theory of relativity has given us a new view of causality and has made room, as it were, for synchronicity. In relativity, the speed of light is just about the only thing that is absolute: it is the same for everyone, regardless of how fast one moves; and it is finite, 300,000 kilometers per second. This is an extremely high velocity. It would enable one to encircle the earth at the equator more than seven times in one second or to travel from New York to San Francisco in less than $^1/_{60}$ of a second. But however great or inconceivable this speed may be, it is *finite*, in contrast to the earlier held belief that light travels at an infinite rate. And relativity tells us that this is the upper limiting speed on all travel and transport of matter and energy. (Thought or an imaginary geometrical point may travel faster than light, but nothing real or physical can.)

This finite speed of light is the key to our new understanding of causality, for two events can have a causal relationship between them only if they are separated in time and space so that a light signal or something slower can travel between them. For example, if I place a phone call in New York and your phone in San Francisco rings exactly $^1/_{100}$ of a second after I finish dialing, it cannot possibly be my call you are receiving in San Francisco, because that would require electrical energy to travel across the United States faster than the speed of light. My call could not possibly reach you any sooner than $^{1}/_{60}$ of a second after I have dialed, for that's the least time it takes an electrical signal to cross the United States according to relativity. As a matter of fact, the phone call would take more than $^{1}/_{60}$ of a second, because electrical signals travel at somewhat less than the speed of light through metal cables, and further delays result from relaying and enhancing the signal in its long journey. But slower than light is okay in relativity. I can leave from New York by car and arrive at your home in San Francisco four or five days later. I can, in fact, initiate any *cause* in New York that produces some *effect* in San Francisco, as long as I don't try to do it in less than $^{1}/_{60}$ of a second, which would be impossible. The fastest thing I can do is to send a direct radio signal (since radio waves also travel at the speed of light) from New York to San Francisco. Relativity places a limit on the time between two causally connected events, one in New York and the other in San Francisco: the West Coast effect must follow the East Coast cause by no less than $^{1}/_{60}$ of a second.

What about the earlier example of the call you receive in San Francisco only $^{1}/_{100}$ of a second after I've finished dialing in New York? Well clearly, that has to be a different call. The two calls are unrelated events, meaning *causally* unrelated. For you to receive my call in $^{1}/_{100}$ of a second or less would be a violation of the theory of relativity and of our present fundamental understanding of causality. A special case of this violation would be for you to receive the call instantaneously. This would have been possible had the old belief been true that light travels at an infinite rate. But because we now know that light travels at a finite speed, two simultaneous events separated in space can never be causally related, since no signal can possibly travel between them in zero time. Under the old belief in an infinite speed of light, it would have been a violation of causality for the call to arrive in San Francisco *before* it had left New York. But according to relativity, it is equally preposterous for the call to arrive instantly or in $^{1}/_{100}$ of a second.

Now suppose you get a hunch about my call and pick up the receiver while I am dialing and then hear my voice. According to relativity, the reception of the signal that you interpret as my voice is causally related to my dialing, but your hunch is not. Relativity distinguishes between two kinds of event-relations. Any pair of events that can be causally connected are said to have a *timelike* separation between them; all other pairs that cannot be causally connected are said to have a *spacelike* separation. My speaking into the receiver and your hearing it are events that are causally related or that have a timelike separation. My dialing and your simultaneous hunch are two events that are not causally related (according to relativity) and that have a spacelike separation between them. Nevertheless, we feel a human symbolic connection between two such coincidental events in spite of the spacelike separation between them. That science cannot give a causal explanation in physical terms, that it does not even allow for the possibility of one, can, in no way, diminish the deep feeling of the gift of communion with another human being or creature that accompanies such intuitions. And it is this feeling of connectedness with other things and people, almost as if by a bestowal of grace, that Jung's concept of synchronicity is getting at.

At the beginning of this section, I said that relativity makes room for synchronicity. This metaphor is more than casual, for there is a vast region of spacetime, consisting of events that have only spacelike or noncausal separations from the immediate here and now. To understand this, one must first recognize that the term *spacetime* means exactly what it says. It is a conception of space and time taken together as a common medium. It is usually imagined as a four-dimensional space in which time is the added dimension on a perfectly equal footing with the three ordinary spatial dimensions. Of course this is only a representation or metaphor, but then, a major theme of this book is that our ordinary conceptions of space and time are equally metaphorical. This construction of a four-dimensional spacetime is so explicit and conscious that relativity gives us the bonus of allowing us to witness the creation of a metaphor for reality right before our eyes.

Naturally, a four-dimensional diagram cannot be drawn in a book, or even imagined very well in our mind. Mathematics is perfectly capable of analyzing higher-dimensional worlds, but our imaginations are quite limited in trying to picture them. Consequently, a lower-dimensional representation is used which is simpler to understand and draw, but which illustrates the essential features of relativistic time and space. One can represent events in time and space on a two-dimensional plot, in which one axis represents space and other time. In Figure 4-1, the horizontal axis represents distance in miles along a one-dimensional continuum, like that of the worm world in our earlier example. The vertical eaxis represents time in seconds. Thousands of miles and fractions of seconds have been used for convenience, but clearly many different scales are possible. Our particular choice is determined by the high speed of light and the desire to represent conveniently the example of the telephone calls we have been discussing. Of course, we don't live in a one-dimensional world, but the essential features of the argument will emerge by reducing,

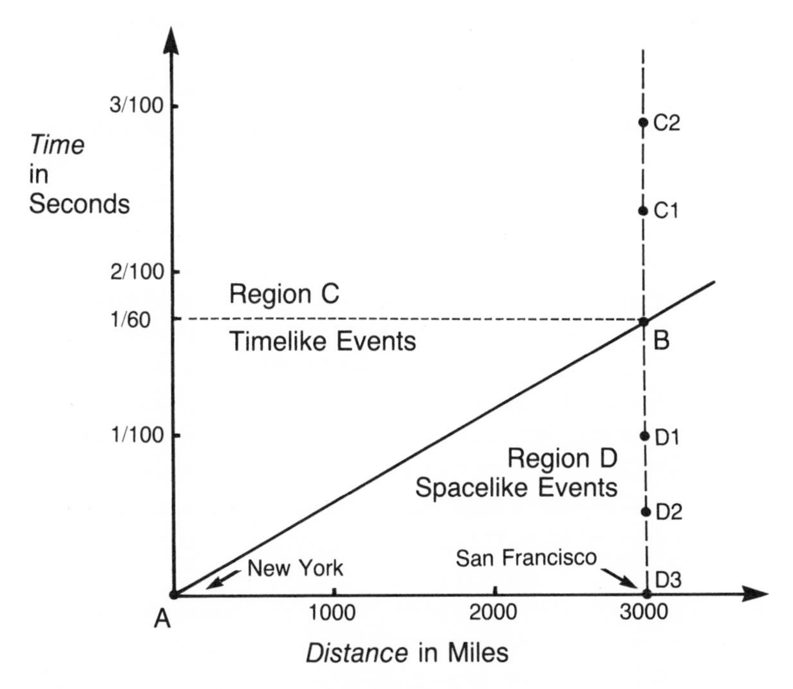

Figure 4-1.
Two-dimensional spacetime

in imagination, the space of our world to just a one-dimensional highway or telephone line from New York to San Francisco.

The point labeled A at the origin of the axes in the diagram represents an event in spacetime. It occurs at time zero and distance zero. Any other point on the diagram also represents an event in spacetime. Here the word *event* is used in a specific sense, but one that is related to common usage. In relativity, *event* means an occurrence at both a specific time and a specific place. In everyday speech, either the location or the time of an event may be rather vague, but in relativity, an event must have a precisely stated time *and* spatial position. The event A in Figure 4-1 represents the instant, in New York City, at which I complete placing my call to you. The time may be 3:00 P.M. or 11:45 A.M. on a certain day of a certain year, but I arbitrarily call it time zero, as if I were resetting a stopwatch. The position is that of my phone in New York, but again, I conveniently label this the zero or starting point on our imagined one-dimensional path from New York to San Francisco.

Event B represents the ringing of your phone in San Francisco. B occurs just about $^{1}/_{60}$ of a second later than A and 3,000 miles from it. It represents the *earliest possible* event in San Francisco that can have a *causal* connection with event A, since light or electrical energy has just enough speed to traverse the distance between events A and B in the time interval that separates them.

Events C1 and C2, which are also 3,000 miles from event A and are separated from it by a time interval of *more than* $^{1}/_{60}$ of a second, may also bear a causal connection to A. The same would be true for event C3 (not on the diagram), much farther up on the dotted vertical San Francisco line, which would represent my arrival in San Francisco by car four or five days later than event A. But events D1, D2, and D3 can have *no* causal connection with A, according to relativity, because they occur at a time and place such that light could never reach them from event A. Light cannot travel 3,000 miles in $^{1}/_{100}$ of a second or less, certainly not in zero time as would be necessary for it to reach event D3. D1, for example, might represent the other unrelated call you got before mine reaches you. D3 would be the hunch you had that I was calling you. All events like B, C1, C2, and C3 are said to have a timelike separation from A and *may* be causally related to it. They need not be: for example, C2 could be an unrelated call. But C2 always has the possibility of a causal relation to A. On the other hand, D1, D2, and D3 have a spacelike separation from A and can under no circumstances bear any causal connection to A.

Similar arguments can be made for events (not shown in the diagram) on a Chicago line, 1,000 miles from A, and on a Denver line, 2,000 miles from A, and so on. Thus all events represented by points on the diagram lying above the slant line connecting A and B have a timelike separation from A and may have a causal connection to A. All points lying below the A-B line represent events with a spacelike separation from A, and they can have no possible causal connection to A. With respect to event A, then, all of spacetime is divided into two distinct regions, C and D. The C region represents future events with a possible causal connection to A, and the D region represents future and simultaneous events with no possible causal connection to A. Region C is timelike; region D is spacelike. The A-B line acts as a boundary between the timelike and the spacelike regions of spacetime. It is a kind of edge of causality. The A-B line is also called the light line, for the points on this line correspond to the events that label the arrival of a light beam that starts from point zero and travels along our one-dimensional space. B, for example, could represent the arrival in San Francisco of the light beam that started from New York at time zero. And since the speed of light is universally constant, all light events initiated by event A must lie on the A-B line. Any event off this line would represent the transmission of something at less than or greater than the speed of light (the latter being impossible for physical matter and energy).

This light boundary between timelike and spacelike events could not have existed if the earlier belief in the infinite speed of light were true. Under such a supposition, the A-B line would coincide with the horizontal axis, since light could traverse any finite distance in zero time, and there would be only one region of space, a timelike one. In terms of our present knowledge, the light line slants upward on the spacetime diagram because light cannot cover any finite distance without the elapse of a finite interval of time; on the diagram, it cannot move horizontally through space without also moving vertically through time.

We can extend our two-dimensional spacetime world (one dimension of space and one of time) to three dimensions (two for space and one for time) by rotating the diagram about the vertical axis as in Figure 4-2. The original horizontal distance axis would then trace out the plane of a two-dimensional world like Flatland. Each of the infinitely many lines radiating horizontally from the center of this plane would form, with the vertical axis, a one-dimensional plot like that in Figure 4-1. Imagine for the moment that the surface of the earth is flat. Each horizontal line from the center of the plane would represent a different intercity line: New York-Juneau, New York-Toronto, New York-London, New York-Dakar, New York-Sydney, New York-Rangoon, and so on. Then the original New York-San Francisco line would be seen as a kind

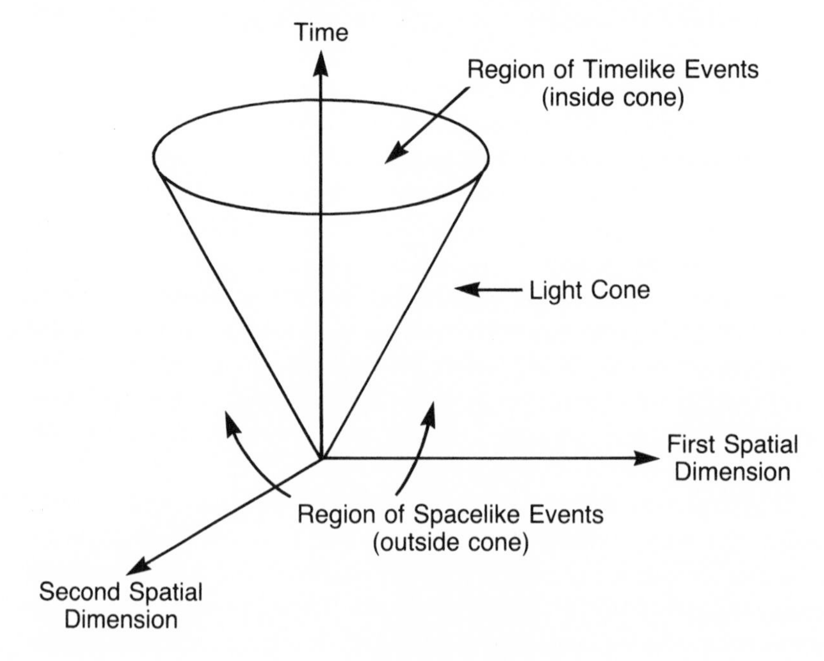

Figure 4-2.
Light cone in three-dimensional spacetime

of special case, one-dimensional world extracted from the two-dimensional plane.

In this rotation, the A-B light line sweeps out a cone, called the light cone, which is illustrated in Figure 4-2. The light cone has become a very famous metaphor in modern physics for the boundary in spacetime between spacelike and timelike events. The term is even used symbolically in talking about the four-dimensional representation for the physical world, although clearly we cannot picture four-dimensional geometry, and, in any case, the bounding surface would not be a cone. (In fact it would be not two- but three-dimensional.)

Thus our knowledge of the constancy and finiteness of the speed of light taken together with the theory of relativity, which adopts this speed as an ultimate limit and gives us a new combined view of space and time, leads us to conceive of the physical world as divided into two regions of spacetime by the light cone. And this division separates the causal and potentially causal from the noncausal, the timelike from the spacelike. The vast region of spacetime lying outside the light cone contains all the events that can have *no* possible causal connection with the immediate here-and-now event at the apex of the cone. Relative to our immediate consciousness at the present instant and location, there are an infinite number of events permeating the infinite reaches of spacetime which are neither available nor communicable to us through any causal means.

We experience only what the light cone, in effect, allows us to experience. The sun which exists at this very instant of time, radiating its brilliance into space, is not the sun that I see or that I can know. The sun that I see is but an image of an event that occurred 8⅓ minutes ago, for that is how long it takes for light to cross 93 million miles that separate sun from earth. If the light of the sun were to be extinguished at this moment, so they tell me, I would know nothing of it for the next 8⅓ minutes. The stars I see in the night sky are not the stars as they presently exist, assuming that they still do exist. They are images of events (emissions of star light) that occurred hundreds or thousands of years ago, depending on their distances from me. The people and objects I see and feel around me are mere shadows of some "real" existence or events that occurred at some time infinitesimally earlier, but earlier nonetheless. When I pull my finger from a burning flame, the finger I pull and the flame I pull it from are not the ones that experienced and perpetrated the damage. I have no thought, perception, or experience that is not the hollow afterimage of a former reality.

We sit here on our little planet, peering through our telescopes, in effect, down that horrendous light cone, seeing nothing but a historical movie. Nothing is real; nothing is present. There is no present. All information is just ephemeral impressions of presumed past events. Individually, we each carry our light cones around with us like funnels through which we must strain and filter the world. This miserly filtering and scrimping light cone is dramatically different from the proverbial horn of plenty with its generous and nourishing bounty. The light cone offers no sustenance and light, no milk and honey, but only a darkened glimpse of paradise, a few distilled drops of the elixir of life.

Does there exist any clearer illustration of the bankruptcy of our modern spacetime metaphor? Not only is there in that realm outside the light cone, in that vast region of synchronicity, a vital and vibrant world filled with all the living events of human imagination, dreams, fancy, myth, intuition; not only is that synchronous spacelike world linked to us through all the dimly perceived channels of knowledge that we in our ignorance vaguely refer to as illumination, revelation, astrology, ESP, clairvoyance, déjà vu, grace; not only has that noncausal world become the dungeon where all human meaning, magic, and madness are held captive; but now, to top it all off, we are to believe that even the physical world that has been redeemed by science from all subjective horrors is eternally forbidden to us. It has no immediate causal presence for us. It is suspended in the remote spacelike realm, and we are left with nothing more real and present than empty dead shadows and phantoms floating through our light cones.

This most potent metaphor of the light cone, however, does more than simply reveal the limitations of a world conceived in purely objective terms. It hints, too, at that territory of human experience and knowledge which communicates to us without the transmission of matter and energy through space and time.

The light cone offers no sustenance and light, no milk and honey.

The space outside the light cone is the space of synchronous events where our hunches, dreams, and wisdom come from In-deed, the space within the light cone is also synchronous, for many pairs of events with a timelike separation still admit of no causal explanation. You in San Francisco might get a hunch about me in New York a full minute after I think of you. Clearly, a physical signal could have been sent from me to you in that time at much less than the speed of light, but it wasn't. The space of synchronicity fills the light cone, both outside and in. Only physically causal space is limited to the interior of the cone, the only region, presumably, with which we have any communication. It is thus the physical notion of causality, which we insist on associating with our concept of linear, sequential, continuous time, that sustains the illusion of our limitations.

Causality and synchronicity are not polar opposites comple-menting one another. Rather, causally related events that can be interpreted only in terms of the modern metaphors of space and time form a subclass of the far more numerous synchronously related events. Causality is thus a metaphor-linked notion. It can exist only relative to a specific view of space and time. Causality depends on the notion of an *extended* spacetime, one possessing distance, interval, measure. It is because it takes a certain time to cross a given distance that a mechanism for causality can even be envisioned. Scientists have been highly imaginative in devising such mechanisms, but this does not change the fact that they all depend on motion in time through extended space.

In the chaotic, nondimensional, totality-point, no causal expla-nation would be needed. Everything *is* everything else, and even the idea of relationship is meaningless. Indeed, the synchronicity principle does not depend on space and time at all. It is only an approximate and suggestive way of hinting at a level of existence and consciousness that is devoid of space and time as we know them. Astrology will never be understood in terms of planetary influences traveling through space. Ultimately, astrology has nothing to do with causality. It is a description on *this* plane of an unfathomable connection on some *other* plane between two things, called astronomy and psychology, that we think of as separate and causally unrelated. And for the most part they *are* causally unrelated. Their connection can only be described as synchronistic as we see it, but it stems from the deeper mysterious connection, even identity, of all things.

Fundamentally, the notions of causality and of a continuous motion in space and time are inseparable, even synonymous. As Dantzig said in the earlier quotation, "These vague ideas of continuity and causality are so closely associated that one is constantly invoked to support the other." And the theory of relativity, by placing an upper limit on the velocity of matter and energy, has made the connection between causality and continuous motion clearer than ever. Since causal information can only be transmitted continuously and at the speed of light or less, the light cone comes into existence along with the distinction between timelike and spacelike separations. If light traveled at an infinite rate, the light cone would flatten out, as it were, to include all space as potentially causal. Even simultaneous events separated in space could communicate with each other if signals could cover a finite distance in zero time. But, of course, this is of no help since all our physical experiments, observations, concepts, and theories rule out infinite magnitudes. So the universality of the motion of light along a continuous path at constant speed gives a unique quality to the light cone and an unalterably limiting character to causality. Because light is in continuous motion at a finite speed, there is no conceivable causal explanation to your premature hunch about my phone call. Only a synchronistic one.

I often think that light is the key to the mystery of spacetime. That which gives us the sight and knowledge of our world is so thoroughly enmeshed in space and time that we can have no vision without them or outside them. In relativity, light plays such a pivotal role in the integration of space and time that it is difficult to think of light as some mere physical phenomenon occurring *in* space and time. It seems almost as if light were the source of space and time. Then again, why should light be more fundamental than anything else? It's just that some mysteries are more appealing than others.

Chapter 5.

What's the Matter?

We turn now to matter, the third of the cardinal metaphors—space, time, matter, and number—on which physics is founded. Using intertia and gravitation as the central illustrations, we shall explore the intuitive and psychological basis of the matter concept as used in physics.

Although it is more intuitive to think of measuring matter in terms of its quantity or substance, it is the practice in physics to use, instead, a more abstract, but more physical, property of matter—inertia. This is the tendency of a material body to remain moving at constant speed along a straight line when no external influences act on it. When an object has been given an initial impetus in free space, for example, it will continue to move indefinitely in one direction at constant speed of its own accord. This tendency, together with the natural resistance of the object to changes in speed or direction, is called the inertia of the body. It takes a force to accelerate an object, but not to keep it moving at constant speed. We shall emphasize the inertial view of matter because it brings out clearly the deep connection between matter and spacetime. But since we are discussing intuitive and subjective approaches to physical concepts, we shall spend at least a little time on the notion of matter as substance.

The point I'd like to emphasize is that modern physics has really pulled the rug out from under our ancient, ingrained notions of matter as some kind of substantial stuff that is impenetrable and indestructible. It was generally believed until the early years of the twentieth century that atoms, of which all matter is assumed to consist, come in the form of hard little balls. This is an interesting example of transferring a metaphor from one level to another: Newton's mechanics had been so successfully used to analyze the solar system in terms of the forces between the planets and the sun that this approach became the prototype for all physical theories. The attack was the same at all levels and size scales whether one thought of planets revolving around an attracting sun, colliding billiard balls, or the tiny molecules of a gas bouncing against each other and their containing walls. Many gross features of matter were derived from a microscopic Newtonian analysis, treating atoms and molecules as tiny particles. It was only natural that people began to believe in the reality of the models. The particle metaphor was transplanted from the macroscopic to the microscopic level, replacing, rather than rejuvenating, the ancient Greek *atomos*. We began to develop a picture of solids, liquids, and gases as mostly empty space, sparsely populated with tiny molecules. But the notion of hard, impenetrable elementary particles of matter has been totally discredited by two major conceptions of modern physics.

First, the so-called elementary particles have lost all their claim to elementarity. An element is supposed to be a *simple* building block in some more complex structure. For example, an endless variety of buildings may be made out of simple bricks. By the same token, all the fantastic forms of matter we know of in the universe consist, presumably, of just the 92 natural elements. But this idea is deceptive, for although the elements may be understood as 92 variations on the basic atom, the core or nucleus of the atom seems composed, as we delve ever more deeply into it, of an unending series of strange, complex structures, which, from sheer force of habit, we continue to call elementary particles. Our attempts to understand the 92 nuclei that hold the basic atoms together have carried us into a world of hundreds of subnuclear particles. Each new candidate, or family of candidates, for the coveted elementary status, invariably turns out not to be simple, but to have an internal structure and to contain within it yet more "elementary" constituents. To expect some kind of ultimate, impenetrable unit of matter to emerge from all of this seems highly unlikely. It is claimed that quarks are the end of the line. Perhaps, but we've heard that story before. Furthermore, there are already thirty-some-odd different kinds of quarks, not to mention the gluons that hold them together, and none of them can ever be isolated for direct observation. Quarks notwithstanding, modern concepts of matter seem neatly summarized in a statement made over fifty years ago by Bertrand Russell:

> The matter that we construct is impenetrable as a result of definition: the matter in a place is all the events that are there, and consequently no other event or piece of matter can be there. This is a tautology, not a physical fact; one might as well argue that London is impenetrable because nobody can live in it except one of its inhabitants.[1]

An alternative approach in modern theoretical physics to the hierarchy of the bewildering elementary particles is the *bootstrap* method. The picturesque idea behind this term of lifting yourself by your own bootstraps lies very close to the fundamental theme of this book, namely that humans are the responsible creators of their world. Bootstrap theories give an equal "democratic" status to all the objects of microphysics and do not grant a more elementary or ultimate standing to any of them. Electrons, protons, and neutrons, for example, in bootstrap theory, are considered to be no more fundamental than atoms of oxygen or uranium. Each implies and is implied by all the others. They lift themselves into existence by their own bootstraps so to speak. Boostrap theory is far less fashionable today in the physics community than the elementary-particle view. But one aspect of the bootstrap approach is intriguing. Pushing the bootstrap hypothesis to its extreme, one is reminded of the ancient idea that each thing is a reflection or an aspect of everything else. I have already argued that this is tantamount to the collapse of the spacetime metaphor altogether, not to mention that of matter. For to unify all things into one undifferentiated whole implies the absence of space and time. When present, space and time provide a basic means of distinguishing among things, by position and age if nothing else. Without space and time there would be no distinction or identity, and any kind of analysis or physical theory becomes almost inconceivable. In the search for the elusive unified theory or single equation that will explain the whole universe, one must be careful not to go too far. Even an equation must have two things to equate. If all things are in some ultimate sense one and the same, as the bootstrap idea seems to suggest, there cannot be any *relationships*, for there are no different *things* to relate. Physical laws expressed in terms of relationships would then be impossible.

But I shall leave such wild speculations, and proceed instead to the second major theme of modern physics that demolishes our idea of matter as substance. The mathematics of quantum theory does not describe particles of matter localized in space as Newton's theory did. Rather, it describes only the probability of finding an electron or a proton at a point in space, never its definitive position. In fact, the impossibility of determining simultaneously the position and velocity of a particle is a fundamental tenet of quantum theory, called the Uncertainty Principle. All predictions in quantum theory are based on probabilities. We can never know exactly where in an atom a particular electron is, but only the relative probabilities of its many possible locations. We can say, for instance, that a certain electron is more likely to be found at position A in an atom than at position B. But we have no guarantee of finding it at A rather than B when we look for it. Even if it is ten times more probable that an electron is at position A than B, there is still one chance in ten of finding the electron at B. The odds are ten to one against it, but it is decidedly possible. If we make a hundred observations, we should find the electron at B about ten times. But we have no way of knowing in advance of any observation whether the electron will be at B or A. Furthermore, we might find the electron at B eight times or perhaps thirteen times out of one hundred. Such are the vagaries of probability.

Now it might seem that modern physics is vague and indefinite at the microscopic level. But the great power of quantum theory lies in its ability to predict and accurately describe phenomena that are observable on the macroscopic scale in the everyday world. Predictions are based on the probabilities of happenings in the minute world of the atom. The method works because of the law of averages, which says that probability predictions become more and more reliable as the number of cases increases. Since gross matter is viewed as consisting of trillions upon trillions of atoms, the individual fluctuations of any one of them is of negligible consequence. If I toss a coin ten times I ought to get five heads and five tails, because heads and tails are equally likely. In practice, I may get four and six, eight and two, even ten and zero. A few peculiar tosses spoil the average. But if I throw the coin a thousand times, I am much more likely to get the predicted 50-50 distribution than in the case of ten tosses. Occasional runs of heads or tails have little effect on the average.

An atom gives off light of a distinct color when an electron within the atom makes a certain characteristic "jump" (transition between orbits). Different colors correspond to different jumps, and different jumps have different probabilities. We cannot predict which particular color of light a given atom will emit, any more than we can predict heads or tails in a given coin toss. But if we have an enormous sample of identical atoms, as, for example, in the case of neon gas enclosed in a glass tube, we can confidently predict that the most likely color of light for a neon atom to emit will in fact dominate the color of the glowing gas. Every advertiser knows with supreme confidence that a neon sign will be red. Quantum theory makes precise macroscopic predictions on the basis of microscopic probabilities. It works far better than Newtonian physics when applied to the atom. Today, the microscopic quantum theory of matter has completely eclipsed the Newtonian theory. But at what price?

Popular expositions of modern physics often talk about the electron in an atom as a cloud rather than a particle. The old planetary model of the atom, in which the electron is pictured as revolving about the nucleus like a planet around the sun, is out, and the new cloud picture is in. We are told that quantum theory portrays the electron today as smeared out in space like a cloud or mist around the nucleus, and we are often shown remarkable pictures of these electronic cloud patterns in the atom. We are now to picture the electron as a fuzzy cloud. Matter is to be visualized as an enormous aggregate of these smeared out clouds, rather than of discrete particles. It is nebulous instead of grainy.

But in truth quantum theory gives us anything but a picture or visualization of matter. The terrible price we've had to pay for ousting Isaac Newton is our visual conception of matter. The cloud pictures do represent something, but not what it says in careless textbooks. The pictures represent what is technically called the *probability distribution* of the electron in the space of the atom. The quantum equations tell us the likelihood of finding an electron at a certain place in an atom, and they tell us this for all possible locations in the atom. An atom is assumed to exist in three-dimensional space, and an electron can be found anywhere in the space around the nucleus. If you picture the nucleus at the center of an imaginary globe marked with circles of longitude and latitude, quantum theory tells us the chances of finding an electron at different distances from the nucleus and at different latitudes and longitudes for each given distance. The probability of finding the electron is different, in general, at each position in space.

Electron cloud pictures attempt to represent all this probability information graphically. Where the clouds are most intense, the probability is highest; where they are most diffuse, it is lowest. But these are not pictures of an electron, they are pictures of *probability.* No one has seen or ever will see probability. It is a mathematical abstraction, and any picture of it is also an abstraction. When you go looking for an electron, you don't find its probability; either you find an electron, or you find nothing. A probability does not properly exist in physical space, and to represent it spatially is simply a convenience to help us picture the mathematics of quantum theory. The probability cloud, or wave as it is often called, which quantum theory *does* describe for us, is not in any sense the same as the electron, which quantum theory does *not* describe. We may foretell the chances of finding an electron at some particular location in space, but we are expressly forbidden by the Uncertainty Principle from using this information to construct any kind of a picture of how an electron exists *in* and travels *through* space. (There are those wonderful prepositions again.)

We cannot, according to the Uncertainty Principle, simultaneously measure the position and velocity of an electron. But knowing the instantaneous position and velocity of a particle and inferring from them, step by step, the future course of its motion is central to Newtonian mechanics and to the Newtonian picture of particles as entities existing in and moving through space. We have given up our picture of matter in space. At the very least, we must admit that if we insist on talking about and imagining electrons with well-defined positions and velocities in space, we are using conceptions that bear no resemblance to our physical theory. In the words of N. R. Hanson, "The impossibility of visualizing ultimate matter is an essential feature of atomic explanation."[2]

Where does all this leave us? We saw that we have no elementary particles, and now we see that we don't have any particles at all, nor, in fact, any kind of matter in space. You may wonder why we even talk about elementary particles and what we mean by their properties such as mass and electrical charge. Well, such terms are a kind of convenience—a shorthand way of referring to a collection of macroscopic measurements that we associate with some abstraction called a particle. No one has ever seen an electron or directly observed its mass or charge. But certain procedures carried out in the large-scale physical world will consistently result in a set of measurements that we associate through a long abstract argument with a thing called an electron. We have no way of visualizing how an unpicturable electron can possess a mass and a charge or how it can be present in space. We use such terms by analogy with our classical Newtonian models of matter, for we have no other pictures to use.

We are left with a theory that provides us with no ultimate unit of matter and no way of even picturing what we mean by matter. We have only the mathematical laws of quantum theory, with a prescription for getting numbers from them. And these numbers will correspond quite accurately, for the most part, with certain macroscopic measurements made in the laboratory. But exactly what it is we are ultimately measuring and how we are to picture and conceive of it is very difficult to say. So much for substance. Could Aristotle have foreseen this, he might well have thrown out of the Academy the entire subject matter.

Inertia, or Fear of Flying

When I was a child, a wonderful H. G. Wells film, *The Man Who Could Work Miracles*, succeeded in capturing and holding my imagination.

A naive and unsuspecting little man, Mr. Fotheringay, played splendidly by Roland Young, is granted the power to work miracles by a playful group of centaur-gods who peer disdainfully down on the earth (George Sanders is characteristically condescending in the scene) while arguing over whether humans can handle more power than they have so far been granted. In a series of comic and pathetic episodes, Fotheringay comes to recognize the broad range of his miraculous powers and transforms himself from a simple downtrodden sales clerk into an arrogant world emperor, who then inadvertently brings about the downfall of the world. The world is mercifully restored to its original state, and although some of the gods depart, convinced that their little "experiment" has justified the worthless contempt in which they hold the insignificant earthlings, one of them, who is fascinated by and sympathetic toward the little creatures (not Sanders, of course), remains behind to increase their powers gradually, convinced that they will rise slowly to the stature of gods.

The film seems to champion Apollonian reason and light over Dionysian nightmare and darkness. But this is certainly not why it appealed to me. Although I came to appreciate the role of reason in science later on, as a youngster I was attracted by the seemingly magical powers of science and by its colossal dealings with the awesome mysteries of the universe. In other words, the Dionysian element. To accompany my dream of being Sinbad with the magic all-seeing-eye, I spun new webs of Faustian fantasy in which I was granted Fotheringay's dazzling ability to work miracles. I vowed that I should not fall into the same traps and, above all, that I should immediately "repeal" that confounding law of nature that ultimately wiped out Fotheringay, *The Law of Inertia*.

Fotheringay's climactic final scene stands out clearly in my mind to this day. At the peak of his power and arrogance, he decrees a convocation of the leaders of all the nations in the world. He miraculously transports the most renowned men and women from the four corners of the earth to hear his words and recognize his supreme authority over them all. Bedouin chieftains appear with their camels, top-hatted Englishmen, Africans, Chinese, Europeans, all amazed by their sudden presence at this great world colloquy. Fotheringay explains the situation and then demands that all the problems of the world be solved on this very day. But he is interrupted by his self-appointed spiritual mentor, Maydig, played superbly as a bumbling religious idealist by the great English character actor Ernest Thesiger. The time is too short, Maydig protests. The sun is already setting, and the few hours remaining in the day could not possibly suffice to carry out Fotheringay's orders.

But Fotheringay will not be put off: there is nothing he cannot do, and he commands that the earth stop spinning on its axis so that the day may last until the solutions are found. Maydig frantically protests that it is impossible. He raises his hand dramatically in a desperate gesture to dissuade Fotheringay. He shrieks out, "The Law of Inertia." But he is too late. The command has already been given, and the earth suddenly ceases its primeval rotation. All are instantly swept off their feet by their inertia: they continue to move although the earth has stopped. In the ensuing chaos, Fotheringay survives, for he had previously made himself invulnerable to physical harm. Out of the swirling whirlwind and debris, he manages to utter his last wish. He wills that the earth be restored to its state just prior to his receiving the miraculous powers, that this time around he not receive them, and that all memory of the present events be erased from everyone's mind. And it comes to pass.

Poor Fotheringay. He could command all nations, the earth itself. He could even reverse time. But the Law of Inertia did him in. He forgot to repeal it.

But we shouldn't be too hard on Mr. Fotheringay. His education was probably not too strong in the physical sciences. Far greater minds than his have forgotten about inertia or even refused to acknowledge its existence in the first place. Aristotle built a whole system of physics in flagrant violation of the law of inertia, and everyone believed it for well over 1,500 years. Not until the middle of the present millenium, when the predecessors of Galileo and Newton began to challenge Aristotle, was there any doubt that things moved at constant velocity under the action of invisible forces, rather than of their own accord, as we now believe. Until Galileo and company taught us otherwise, it was obvious to anyone who had eyes to see that an object tended to slow down unless something—some force or agent—kept it going. The accelerated motion of objects falling toward the earth was no contradiction for Aristotle, since it was also true that all matter naturally sought out its proper abode and resting place in the earth. All matter tends downward; all spirit up. And humans are caught in between. A lovely system, really; it had a lot going for it, until Galileo liquidated it.

This idea of inertia was indeed something new in the world—something imagined for the first time. It was a construct, a metaphor, based not in reality but in the burgeoning conceptions of time, space, and motion. Owen Barfield describes it well:

> A very long step—and a very difficult one—was taken in the final ousting of participation, when the Aristotelian and medieval doctrine that all bodies come to rest, unless they are kept in motion by a "mover," was at last abandoned. Yet if we base our hypotheses on the behaviour of the bodies we actually see in motion, this is the only conclusion we can possibly come to. The theory of "impetus," which later developed into the concept of "inertia," requires, not observation, but the abstract, geometrizing supposition, never realized in practice—at least on earth—of bodies moving through a gravity-free, frictionless vacuum. In this case therefore the change of outlook—and there could hardly be a more significant one—must have been hindered rather than helped by observation.[3]

Participation is Barfield's term, derived from Thomas Aquinas, for an older form of consciousness, less geometrical and abstract than ours, in which people felt a participating connection with all things around them. It involves the sense of space that I spoke of in connection with astrology, if it can be called a sense at all. For it is we moderns who have taken all the felt relations of the medieval and ancient mind, reduced them to principles and geometry, and embodied them outside our heads in a cold, alien, empty space. Galileo and his predecessors did this too with our intuitive notions of matter in motion when they invented inertia.

What is this strange law of inertia that confounded Fotheringay's miracles and which is the cornerstone of the whole Newtonian edifice? The law of inertia states in essence that a material body, free from external influences, tends to maintain its state of uniform motion indefinitely. By state of motion is meant the combined speed and direction of the body. Uniform refers to the constancy of these two. Thus, the law tells us that a body moving at this moment in empty space will continue to move at its present rate of speed in a straight line forever. Unless, of course, some external force or influence acts on the body and changes its state of motion. But in the absence of such forces, every material body has this inertial tendency to move at constant speed along a straight path forever. So the psychic mover which Aristotle presupposed (or perhaps, sensed), and which kept things in motion, is now replaced by nothing more causal than a mere tendency: a body *tends* to maintain its state of uniform motion. No explanation, no causative agent: bodies simply tend to move forever. And this we call a law. Indeed, to talk of repealing it is not so preposterous when one thinks how arbitrarily it seems to have been "passed."

I suppose this should really not be too surprising, for all physical law lacks any kind of causative explanation. I don't mean that physical law isn't causal. It is. Gravity is the cause of the particular motion followed by the planets; electron jumps are the cause of the light energy emitted by atoms. And so forth. But what, for example, is the cause of gravity? Newton himself spent many years wrestling with this question. He was reluctant to advocate a picture of the world in which masses seem mysteriously to transmit their attractive influences on each other across the vast reaches of empty space. Nor could he understand what the nature of this gravitational force was and what caused it. He finally persuaded himself to publish his revolutionary new synthesis of the world when he became convinced of its remarkable accuracy in predicting the motions of the heavenly bodies. But perplexing philosophical questions continued to plague him. Newton himself continued to think and write about the whys, but he usually avoided such matters in his purely scientific writings. Until very recent times his nonscientific writings were largely ignored. In today's science we always ask how, never why.

The accepted modern viewpoint in science has been to look upon physical law as a realistic description of how things behave and not as an explanation or justification of why they act as they do. In quantum physics, where, as we have seen, there can be no correspondence between mathematical law and picturable reality, the theory itself is taken as a kind of ultimate rational structure of the universe. In other words, physical phenomena are not conceived in terms of material bodies—substances—mechanistically affecting each other, but as manifestations of pure mathematics. Physical law, itself, is taken as an ultimate. It is as if the equations of theoretical physics were chiseled into the rock of ages or branded into the flesh of space and time, and we need not ask for deeper explanations.

But the twentieth-century tendency to talk of a rational order in the universe or of a comprehensible intellectual structure is revealing. Where in the universe does the rational mathematics reside? Who makes the judgment that it is rational? Whose reason do we see when we peer out there into space or down inside the atom? The purely mathematical view of modern science, which has replaced the mechanistic view of Newton, is forcing us ever closer to the untenable position that the physical world is our own projection. Material mechanisms and machines we may imagine existing in space, but mathematics, equations? The difficulty stems from our spatial metaphor. We picture space as something outside ourselves, unrelated to us, nonhuman. Now that we find in external alien space evidence of mind and reason, we are at a loss to explain it. We are like children unable to comprehend our own images in the mirror we have created.

Inertia is the earmark of our bankrupt spatial metaphor. With gravity and the other forces of nature, one may at least nurse the hope of finding a deeper explanation in terms of more fundamental concepts. (Einstein actually did this in the case of gravity, and we shall consider the nature and implications of his new metaphor a little later.) But the inertial character of a material object, its tendency to move uniformly forever, is something inherent in itself, in the very nature of motion and space and time, with no deeper cause possible. Now one special but very important case of uniform motion is rest. A body at rest has the constant velocity required by the law of inertia, namely, zero. Thus the law of inertia tells us that it is the *natural tendency* of an object to remain perfectly still floating in empty space forever and ever. We are offered no further security or understanding than the mere statement of this law.

Picture, if you can, some unsuspecting primitive man, resurrected from his ancient grave and confronted with this incredible idea. This miserable creature is so insecure as to be addicted to a picture of a world securely resting on the backs of four tortoises. He would shrink in horror from the Greek and late medieval notion of a round world, even though it be placed reassuringly at the center of the universe, drawing all material things safely to its bosom. The displacement of the center of the universe from earth to sun, with us now whirling at dizzying heights from the foundation of things, held in place only by the tenuous chain of gravity, would force the poor fellow to his hands and knees, quaking in a paroxysm of insecurity and pleading for mercy. Now tell him that the whole solar system and all the island star-galaxies of the universe are simply floating of their own accord in empty space with nothing to hold them up but the law of inertia, and he will simply pass away on the spot. And may he rest in peace.

Objects float in space because it's their tendency to do so. Inertia is no causal law, but an ex post facto apology. It's like saying that fire burns because it's hot or that arsenic kills because it's poison. The law of inertia is an arbitrary and contrived excuse for our stability and security in the space we picture as patently tenuous and unsupportive. It is the inevitable psychological anchor for humans afloat on spaceship earth. In the chaotic, unified all-point, no question of floating or even of stability can arise, for there are no different parts requiring support from each other. But once the point is expanded into space and all our supporting connections with the varied articulated parts are severed, we too should collapse in a paroxysm of insecurity without some assurance that we shall not fall helplessly down into the jaws of hell.

Held in place only by the tenuous chain of gravity.

By now we have become so accustomed to our secure lives on a floating world that we have canonized this floatingness into a law of the universe, instead of questioning the spatial metaphor that requires such bizarre extravagances. Blessed be the Law, our God, King of the universe, which keeps us floating in space with no visible means of support. Amen.

Einstein's Metaphor

We cannot contemplate the subject of inertia and matter without dwelling a little on the sublime synthesis of space, time, gravity, and inertia which is the general theory of relativity. Einstein's theory is often maligned as being incomprehensible and tortuous, but I believe that the basic ideas of general relativity are simple enough to teach to elementary school children. No abstract mathematics is needed, nor even much knowledge of physics, but only a little common-sense geometrical intuition. In fact, general relativity is essentially a generalization of the law of inertia, which is why it interests us at this point.

Until now, I have emphasized the view of matter as inertia, as opposed to the view that it is substance, because the inertial property enables us to measure and define matter in the most physical manner, directly in terms of space and time. There is, however, another important property of matter, recognized as a universal characteristic since the time of Newton. This is the gravitational attractive force that all material bodies exert on each other. Every body in the universe, according to Newton's famous law of gravity, exhibits a gravitational attractive force in proportion to the quantity of its mass. The more massive an object, the greater its gravitational pull on other bodies. The gravitational force also diminishes with the square of the distance from the attracting body. If, for example, there are two identical comets in the gravitational field of the sun, one twice as far from the sun as the other, then the more distant comet feels one-fourth as much gravitational pull from the sun as does the nearer comet. But at a given distance from a material body, its gravitational attraction is directly proportional to its mass. At equal distances from the moon and the earth, the gravitational field of the moon is considerably weaker than that of the earth since the moon is considerably less massive than the earth.

What puzzled and intrigued Einstein about these well-known facts was that the force of gravity should be proportional to the same mass that is a measure of inertia. The mass of a body characterizes not only its inertial resistance to changes in motion, but also its gravitational attraction of other material bodies. Why should two such apparently unrelated physical characteristics of a body (resistance to changes in motion and gravitational attraction) both depend on the same property of matter, its mass? It was too much of a coincidence for Einstein. He felt that it could not be a mere accident. No other basic physical force depends upon the mass of an object. For example, the electric field of an electron has the same strength as that of a proton, although a proton is about 1,800 times as massive as an electron. (The force of an electron on another charge is in the opposite direction to that of a proton, but the strength of the two forces is the same.) It is the electrical charge, not the mass, of an electron that determines the strength of its electric field, and electric charge is in no way related to mass. In other words, the inertial properties of an electric charge are completely unrelated to its electrical properties. How hard or easy it is to accelerate it has nothing to do with its electrical attraction. This is true for all electrified objects. It is also true for the magnetic force and the nuclear forces. But the gravitational force, by contrast, seems linked somehow to inertia. But how?

In the effort to answer the inertia/gravity question, Einstein achieved a brilliant synthesis. He made a profound physical interpretation of what had formerly been considered a mere coincidence, revolutionizing and uniting our conceptions of space, time, and matter. In the special theory of relativity, Einstein had fashioned an alloy of space and time, and, in the general theory, he added matter into the melting pot. Einstein observed that under certain conditions it is impossible to distinguish physically between gravitation and inertia: If you are in a sealed elevator falling freely in the gravitational field of the earth, all the objects in the elevator fall with you at the same rate and appear to you to be at rest. There is no way that you could distinguish these free-fall observations from the situation in which the elevator is stationary, somewhere in outer space in the absence of a gravitational field. All observations inside either elevator would be the same. This is true because all objects fall at the same rate in a given gravitational field. We don't usually observe this at the surface of the earth where air resistance slows down feathers, leaves, and parachutes. But as Galileo already knew, when air resistance can be minimized, or at least made the same for different objects, the rate of fall of an object is independent of its mass or weight. Galileo, in the classical legend, simultaneously dropped an iron and wooden ball of the same size from the top of the Leaning Tower of Pisa, and the two reached the ground at the same time. An equally graphic demonstration is routinely done in today's introductory physics classes when a penny is seen to fall at the same rate as a feather in an evacuated glass tube.

The equal rate of fall of all objects in the gravitational field of the earth is a manifestation of the strange coincidence between inertia and gravity. Although the earth pulls harder on a more massive body, such a body has a greater tendency to resist speeding up because it has a greater inertia. The two effects equalize, so that all bodies fall at the same rate regardless of their masses. Objects moving about in an elevator that falls freely in a gravitational field behave in exactly the same way if, instead, the elevator is stationary in gravity-free space.

But this isn't the only ambiguity, Einstein argued. Suppose you are in an accelerated elevator in free space, one which speeds up at a constant rate. Imagine, for instance, that your elevator is being towed by an accelerating rocketship. You no longer float around as in a stationary box in free space. You are now held "down," able to stand on the "floor" of the elevator (the side opposite the towline connection). Your inertial tendency to resist changes in your speed seems to press you against the floor of the continually accelerated elevator. Actually nothing is pressing on you. The elevator is speeding up and you are lagging behind, because like all obedient subjects of the law of inertia, you resist the speeding-up process. Since your feet are on the floor, you do keep up with the accelerating elevator, but your inertial tendency to lag behind makes the accelerating floor push constantly "upward" on you. Inside the elevator, you feel this as a kind of "downward" pressure throughout your body in exactly the same way that you experience the force of gravity as your weight when you are stationary on the earth.

If you hold an object, a key, say, in your hand, it too accelerates along with everything else in the elevator. The force necessary to accelerate the key is transmitted to it from the rocket through the towline, to the elevator and your body. But if you release the key from your hand, what happens then? There is no longer the force of your hand to accelerate it, and it is in free space in the absence of gravity. Now the key, too, is a faithful subject of the law of inertia, and since there is no longer any force acting on it, it will maintain a uniform state of motion. In other words, it will continue to move at the constant speed it had just as you released it. So as you accelerate and speed up, the key continues to move at its slower constant speed and will appear to you to be "falling downward" at an ever-increasing rate. When the key reaches the floor, it will once again be accelerated by the floor pushing "up" on it, and it now appears to you to be at rest. But a stationary cosmonaut in the space outside the elevator would observe (if it were possible to see into your elevator) both you and the key being accelerated at the same rate. To you inside the elevator, the situation is indistinguishable from the case in which your elevator is at rest on the surface of the earth and you have simply dropped a key from your hand to the floor. In fact, any experiment you might perform inside the accelerated elevator produces a result no different when performed in an elevator at rest in a gravitational field. Einstein concluded that uniform acceleration and gravity are physically indistinguishable, and this new *principle of equivalence* became the cornerstone of the general theory of relativity.

Another way of stating the principle of equivalence is to say that gravitational mass and inertial mass are identical. As long as one accepted as pure coincidence that both the gravitational and inertial properties of a body are directly proportional to its mass, it was natural to assume that the body had two different but equal masses. This is by analogy with an electron whose electric field is caused by an electric charge, and this charge is associated with the electrical properties of the electron and nothing else. So it was assumed that there must exist a gravitational charge or mass, distinct from the inertial mass, which caused the gravitational field. But all experimental attempts to find any distinction whatsoever between gravitational and inertial mass have failed. This is in agreement with the principle of equivalence. Indeed, Einstein's interpretation eliminates the coincidence altogether by viewing gravity and inertia as one and the same phenomenon.

In the Newtonian world view, the law of inertia expressed the natural tendency of a material object to move at constant speed along a straight line, or to remain at rest. Any departures from uniform motion, any accelerations or changes in direction, were explained in terms of forces, seen as external agents overcoming the natural inertial tendencies of an object. Thus hypothetical forces are always associated with accelerated or nonuniform motion. If, for example, Mars were all alone in space, or, at least, very far from the influence of other bodies, it would move at some constant speed in a straight line. This is what the law of inertia predicts, and we all believe the law of inertia. But this is not what we observe Mars doing. We see Mars traveling along an elliptical orbit at a varying speed. As Mars moves, it constantly departs from the path that the law of inertia says it must follow. How are we to reconcile this departure with our faith in the law? Well, the answer is very simple. We need only postulate the existence of a demon or psychic mover, called gravity, who is, of course, invisible and undetectable, and who manifests himself only by producing the required departures from the law of inertia. Mars would like nothing better than to proceed lazily with its uniform motion, but gravity (pardon the small *g*, but a capital letter might reveal that we have actually christened a creature of our own creation), that little demon, never gives Mars a minute's peace. It keeps on prodding him to disobey the law of inertia and follow that diabolical ellipse.

"Rubbish," says Einstein. "We don't need any demons to get us out of this mess. We need only realize that all the moving objects in the universe aren't departing from the law of inertia; they are obeying it. The motion of all these objects isn't crooked, space is! Things speed up and slow down and follow curving paths because the spacetime continuum we inhabit is curved, not flat as Euclid thought. All motion really is uniform, but since spacetime isn't flat, uniform motion isn't necessarily at constant speeds and in straight lines. Objects do follow the straightest possible lines, if by a straight line we mean the shortest distance between two points."

On the surface of the earth, the shortest distance between London and Tokyo is not a straight line. There are no straight lines on the surface of the earth. The shortest distance between any two points on the surface of a sphere is a great circle. If the spacetime we live in were only two-dimensional and were curved like the surface of a sphere, all objects would naturally tend to move on great circles and that's what the law of inertia would say. The creatures in the plane of Flatland obey the old-fashioned Galilean law of inertia, since the shortest distance between points in a flat two-dimensional plane is a straight line. The straight line is just a special case that occurs in flat geometry. The path that is the shortest distance between two points in any arbitrary space that is not necessarily flat is called a *geodesic*. The geodesics on the surface of a sphere are great circles, and in a plane, they are straight lines. For creatures living in other two-dimensional surfaces, like that of a trumpet or a saddle, the geodesics would be other less familiar curves, and their law of inertia would require that bodies move along their geodesics.

The terms flat and curved have an intuitively understandable meaning when applied to two-dimensional surfaces, like planes, spheres, and trumpets. But we don't live in a two-dimensional world or even in a three-dimensional one, as we had always thought. In the relativistic view, time is added to space to form a four-dimensional spacetime continuum. The terms *flat* and *curved* are used only in analogy with the two-dimensional case, since we cannot picture four dimensions, let alone a four-dimensional curved space. (We can picture a two-dimensional space as curved in the third dimension. To picture a curved four-dimensional space, one would have to begin from the viewpoint of five dimensions!)

Now we are in a position to appreciate Einstein's grand synthesis. Gravity hasn't really been united with inertia, it's been dispensed with altogether. We no longer need to introduce some mysterious force of gravity to explain Mars's departure from the law of inertia. Mars does obey the new generalized law of inertia in our curved spacetime, which says that all bodies follow the geodesics in their local region of spacetime. Mars traces out an ellipse at varying speed because the geodesics in our local solar system spacetime are shaped that way. (It is time together with space that is curved, so that both the shape of a trajectory, as well as the speed along it, are determined by the geodesics in four-dimensional space. If you can't picture this, don't worry; you're not alone.) In Einstein's picture, all motion is inertial; there are no departure motions. Our two elevator examples could not be distinguished from each other because they are both inertial. All bodies tend to follow geodesics in spacetime. The apparently irregular motion of an object, which was formerly rationalized by invoking an elusive force of gravity, acting mysteriously across enormous distances, is now seen simply as a symptom of the peculiar geometrical character of our curved four-dimensional spacetime. Physics has been reduced to geometry.

There is, however, a hitch in this grand design. How is it that spacetime becomes curved in the first place? The planets are seen to obey the law of inertia in a curved spacetime, but why is the local geometry such as to produce ellipses? Can we explain the peculiar geometry of our relativistic solar system any more easily than we could explain the peculiar gravitational field of our classical one? Unfortunately not. The sun, once again, is the culprit. It is the cause of the geometry which in turn is characterized by the geodesics necessary to explain the inertial motion of the planets. Inertia is indeed geometry, but geometry is yet traceable to an independent source in the form of matter. It is the mass-energy content of space which determines its geometry—its inertial character. The more concentrated is the mass and energy in a certain region of space, the greater is the curvature of that region and the more curved are the geodesics. A big star has "tighter" planetary orbits than a small one because the space in its locale is more tightly curved. Material bodies in space are still affected by the size of other material bodies in space. Geometry has become a new name for gravitational field.

Despite these shortcomings, the general theory of relativity is, to my mind, the most magnificent edifice among the towering architectural triumphs of physics. It comes tantalizingly close to identifying matter with space and time themselves. Eliminating gravity and reducing inertia to geometry is an incredible leap, and the structure of the theory will delight any musician or architect who can interpret its mathematics. Furthermore, general relativity gives a more accurate description of both the solar system and the universe at large than does the Newtonian theory it displaces. But I believe it founders philosophically on the same treacherous reef that has shipwrecked many a fine metaphor before it. It continues to assume a vast empty space, external to the subjective world of the human mind. Within such a conception there can never be a resolution of the unyielding paradox of a space, somehow synonymous with matter and yet containing it. The basic equation of general relativity equates the mass-energy distribution of a region of spacetime with its geometry. But if space is mass, how can it contain mass or how can the mass be bunched up and more concentrated in some parts of space than in others? General relativity seems to be equating two metaphors for our world, mass-energy with spacetime. They are both spatial metaphors with the same geometry.

I welcome this reduction of two metaphors to one. Yet they are somehow treated as different. A planet moves according to the geometry of a spacetime determined by a sun which is like a lump in that spacetime. Is that lump part of the spacetime itself, or is it a foreign element in it? If the lump is pure geometry, why is it there at all? Why should space be tighter, so to speak, and more concentrated in some places than in others? If there is nothing but pure space, why isn't it all homogeneous? On the other hand, if we must invoke the presence of matter *in* space to explain certain geometrical singularities and regions of high curvature, in what sense are geometry and inertia equivalent to matter? We are trying to hold mass and space apart while bringing them together. In the ultimate all-point, of course, the problem does not arise. Mass, space, time, and everything else are one and the same; there is no extension and no distinction. But with an external articulating space we cannot have things the same and different at the same time. In our space, two things cannot occupy the same place at the same time, not even space itself. General relativity is struggling with the attempt to fuse space and matter, but it cannot without coalescing all things—inside and outside—and collapsing space and time to zero-infinity. Is any purely physical theory or metaphor capable of such a feat?

Body and Soul

When I finished writing the previous section on general relativity, I had the rather uneasy feeling that I had left some loose ends hanging. Try as I might, I could not decipher my feelings and I resolved to go on. But then a remarkable synchrony occurred which provided the missing clues. I had decided to consult the *I Ching*, not for any reason connected with this book, but to get some ideas and counsel about the coming week. The hexagram I obtained was *Yu*, which in the Wilhelm-Baynes translation of the *I Ching* is rendered as Enthusiasm. The English text reads, in part, as follows:

> The inviolability of natural law rests on (the) principle of movement along the line of least resistance. These laws are not forces external to things but represent the harmony of movement immanent in them. That is why the celestial bodies do not deviate from their orbits and why all events in nature occur with fixed regularity.[4]

I felt this was another cardinal at my window which I could not afford to ignore. Einstein himself could not have made a neater summary of relativity. It's all there: the generalized principle of inertia, the absence of external gravitational forces, and the geodesic orbits of the planets. But there is something more, something which general relativity shares with a certain very general principle in modern physical theory. It is *Hamilton's principle of least action*, and the connection with "least resistance" in the *I Ching* reading is obvious and remarkable. The least action approach leads to a mathematical formulation of physical theory which brings out, as earlier formulations do not, the deep connection between apparently different branches of physics, such as mechanics, electricity, light, and heat. The principle of least action is found to underlie, for example, the movement of objects and also the behavior of electric and magnetic fields. It is an integrating principle of great generality in physics.

To illustrate the power and generality of Hamilton's principle, we shall look briefly at how it is applied to optics. The conventional laws describing the transmission of light say that light rays follow straight-line paths in a vacuum or in a transparent medium, like air or glass. Separate laws describe how light bounces from reflecting surfaces and how it bends, or is refracted, on passing from one transparent medium to another—air into glass, for example. So we have different laws for how light behaves *in* a medium and *at* its bounding surfaces. But in the Hamilton formulation, the transmission, reflection, and refraction of light are all described by the single principle of least action which says that between any two points light will always follow the path that takes the shortest time. For example, since light travels more slowly in water than in air, a flashlight beam will bend on entering a lake so that the path in the air, where it has a higher speed, will be maximized and the path in water minimized. The bent path requires the shortest time between given points in the air and water, even though a straight line connecting them might appear shorter.

Now another idea behind the principle of least action which the *I Ching* brings out very clearly is that the natural laws that things obey derive, in some sense, from "the harmony of movement immanent in them," rather than from external forces. For instance, the least action formulation of mechanics attributes the motion of a planet more to its own inherent nature than to an external agent (the sun), as does the Newtonian formulation. And general relativity as well as the law of inertia it generalizes ascribe the natural motion of a body to *itself* or, at the very least, to its *local* region of spacetime, and not to any external agent. It is the local curvature of space, not a distant sun, which causes Mars to follow its elliptical orbit.

The idea of something resonating to its own immanent harmonies harks back to the nonspatial metaphors of the ancients which view all things as aspects or reflections of each other, and even farther back to the notion of a unit-point in which all is one —a microcosm. How does a planet sniff out the geodesics in space it must follow? Does it already know them in its heart? Does spacetime whisper its secrets to Mars as it hurls by? Is the local region already aware of the presence of the sun far away? Does each tiny neighborhood of space already contain all the information and knowledge of the vast regions around it? Is not the relativity metaphor struggling to express through its mathematics what is impossible to say in rational language—that all of space is contained in its every part, the macrocosm in the microcosm; that extended articulated space is an illusion, an idol, which even physics at last must explode?

There is a supreme irony in all of this. For our long discussion of mass as inertia and of general relativity has led us back to the previously rejected, unphysical notion of mass as substance. We first equated mass with inertia and then inertia with geometry in the relativity metaphor. There is a profound and irreducible confusion between mass *in* space and mass *as* space. Space is inseparable from its mass content, and we no longer have a description of space as empty. Space is substance; substance is space. The old idea of a *luminiferous ether,* long ago rejected by the triumphs of relativity theory, has returned to haunt us once again. This ether, although never detected, had been assumed to permeate the whole universe, like a kind of extremely rarefied transparent fluid, to explain how light could travel through space like a wave. But Einstein had explained the behavior of light without the assumption of an ether, and it had been tossed into the waste bin of theoretical physics. Now it seems that space itself is like a material substance whose curved geometry describes for us its grain and contour, or perhaps its blood vessels and sinews.

Of course, there is a marked difference from the case of the luminiferous ether. The spacetime substance of general relativity is highly abstract and mathematical, not a material substance at all. But then all of modern theoretical physics seems to consist of mathematical abstractions whose only connection with concrete reality is numerical prediction. The predicted numbers may be verified by following certain empirical procedures that are tacked on to the laws. In effect the laws say, "Do not ask what the symbols in my equation mean, but if you perform the measurements

I prescribe, your answers will confirm my predictions." The whole universe is like a black box with dials and meters on its outside. We have found some equations that can predict the meter readings, but what connection these equations might have to the unobservable contents of the box remains a mystery. Our modern spacetime ether is but a mathematical artifice, a model for thinking about and describing the world, but having nothing to do with its physical reality. In fact, there's a serious question as to whether physics describes anything *physical* at all.

Modern spacetime is very reminiscent of the *causal body* in the Hindu spiritual hierarchy. This is a plane of consciousness or being above the ordinary physical plane. As with other profound primitive concepts, it springs from a state of consciousness and experience very different from our own, one that is almost nonspatial. For the causal body includes all of what we today should call space and time, past, present, and future. It contains the historic record of all individuals and cultures. It holds the *akashic* record of all past incarnations and acts like a sort of cosmic mediator of past and future karma. In other words, it is a realm of all human knowledge and divine wisdom, not of empty space. It has an organic, living quality, so that causal body rather than something like causal plane is a particularly apt term for it. The causal body permeates space and time and forms a continuum with all human beings, creatures, and things. We all participate in the causal body at some subconscious or, better, superconscious level.

To speak of such an all-pervasive, nonphysical medium is not too different from describing a four-dimensional spacetime geometry. The contours of such a geometry include events that to us would appear as past and future as well as present, since time is part of this continuum. And because this insubstantial spacetime is also a representation of matter itself, it is like an immaterial body. How different is a spiritual causal body from an abstract four-dimensional ether? In a world in which space, motion, and matter are distinct from mind and wisdom, perhaps the closest we can come to a spiritual realm is a mathematical one. The spirit/matter dichotomy slowly undermines itself. Quantum theory and relativity do not provide any substantial, impenetrable, ultimate matter to rationalize for us the tables we pound on and the planets we land on. Matter and substance are mathematical abstractions. Physics tells us that mind and body are one. The causal body and spacetime are metaphors from different cultures that bespeak the unfathomable unity.

Chapter 6.

The Numbers Racket

The fourth of our cardinal metaphors, number, has played a basic role throughout the history of science and cosmology. Pythagoras took number to be the very stuff out of which the cosmos was made, and modern theoretical physics, with its basis in pure mathematics, might seem almost Pythagorean in its outlook. But the modern physicist is primarily concerned with quantitative description, in contrast with the Pythagorean interest in qualitative and symbolic meaning. I place great emphasis on number in this book because, through its connection with measurement and quantification, it has persuaded us more strongly than anything else to trust in and idolize the objectivity of physical science.

Number is as fundamental as the other three cardinal metaphors, space, time, and matter, because it is an interrelated aspect of the divide-and-conquer metaphor which extends and diversifies the primal unity. Number may be the most obvious characteristic of the plurality of our experience, but it cannot really be separated from extension, order, sequence, causality, identity, and other kindred characteristics of our physical world construct. When we see a group of objects on a table, we may give immediate expression to their distinctness and multiplicity by counting them. But we cannot do this unless we perceive the objects in some order, as nonoverlapping, uniquely identifiable things, which assumes a spatiotemporal and material view of the world.

The inevitable fusion of the space, time, and matter metaphors has become more and more apparent in the last few chapters, and we shall now see that number, as well, must be blended into the amalgam.

Dynamic Duality

In the physical sciences, it is primarily the quantitative aspect of numbers that is of interest. Numbers are used to signify the sizes of things. When we say that a table is three meters long, we mean that it is three times longer than the standard of length, the meter. This is really a comparison or a ratio—the ratio of the length of the table to that of the standard meter. A measurement of length involves both the cardinal and ordinal aspects of numbers: for, whereas cardinal numbers designate size or amount and ordinal numbers express rank or sequential order, a measurement implies both. We specify that a table is three meters long by locating its length in the ordered sequence: 1 meter, 2 meters, 3 meters, 4 meters, and so on. We shall see both the cardinal and ordinal aspects of number reflected as well, in their metaphorical character.

We begin with a related aspect of number—its associative, symbolic character. It is perhaps most clearly illustrated when numbers are used simply as labels, as for example in the Dewey decimal library classification system, in which one number is uniquely associated with one book. In science, numbers are often used in this way. For instance, quantum mechanics allows only for the existence of certain stable states of a hydrogen atom. Each of these states is labeled with a unique set of quantum numbers. Such Dewey decimal and quantum numbers serve as surrogate names which may or may not have any meaning. In the Dewey system, the numbers are arbitrarily assigned to different areas of knowledge. In the hydrogen atom, the quantum numbers correspond to certain physical quantities descriptive of a stable state.

Numbers are used as names, or symbols with characteristic meaning of their own, in many other areas. In superstition, for instance, numbers are thought to be lucky and unlucky. The number 13 seems to have such ominous meaning for us that we refuse to employ it as a proper ordinal number in labeling the floors of many apartment buildings. Many people will not live on the thirteenth floor or do anything of importance on Friday, the thirteenth. Few such people know what 13 means, although they recognize that it has meaning enough.

Pythagoras associated definite meanings with the numbers: *one* stood for reason and the primal atom, *two* for opinion, and *four* for justice. Down through the ages, many meanings and magical powers have been associated with numbers. All the divinatory sciences depend strongly on this symbolism. The Sabine symbols of astrology attribute an allegoric image to each of the 360 degrees in the circle of the zodiac. The *I Ching* consists of 64 symbolic hexagrams, chosen numerically by a procedure involving the manipulation of yarrow stalks or the tossing of coins. The hexagrams are based on an ancient system, using binary, or base two, numbers.

These so-called unscientific and nonquantitative aspects of numbers are tied up with archetypal imagery, buried deep in the human mind, and perpetuated in dreams and myth. A profound discussion of this appears in the book *Number and Time* by the Jungian scholar Marie-Louise von Franz. The subtitle of the book is *Reflections Leading toward a Unification of Depth Psychology and Physics* so that its relevance to this book is great, as is the debt I owe von Franz.

In her book, von Franz extends and elaborates on the work of Jung, who, like Pythagoras, saw number as truly fundamental. Jung said, "Number is the most fitting instrument our mind can utilize for the understanding of order."[1] By order, Jung means not only sequential order, but any kind of harmonious pattern recognized by the human mind. We see here Jung's kinship with Pythagoras, for it was the numerical, harmonic order that he found in music and in the cosmos that moved Pythagoras to assign so cardinal a role to number.

Von Franz bases her case for unification on the fundamental connection between the physical and the psychic found in the deep substructure of the mind, Jung's collective unconscious, which underlies both. With Jung she defines number psychologically as an "archetype of order which has become conscious."[2] Number, in this view, is a property of mind *and* matter, expressing our fundamental awareness of the order common to both. Von Franz capitalizes on the qualitative character of numbers we have been discussing, drawing physical examples from the world of atoms and crystals and psychic examples from the realm of dreams and myth. The references to regular crystals and symmetric solids again harks back to a Pythagorean vision of the world in geometric and crystalline terms.

In separate chapters von Franz explores the meaning of the four cardinal numbers a la Pythagoras: *One* is the symbol of the ultimate unity, the *unus mundus*, the primal fount of all things. *Two* is duality, the first mysterious splitting of oneness into distinguishable parts, that incomprehensible articulation which symbolizes the very process of creating an extended differentiable universe out of the primal chaos. *Three* is the tension between one and two, the dynamic actualization of the psychic and physical realms. Three is manifestation. (The triplet pattern in both the *I Ching* and the DNA molecule of the genetic code and the connection each of these has with manifestation or coming into being provide dramatic examples of von Franz's thesis.) But three is also the trinity, the synthesis of one and two, and thus the symbolic return to unity. This retrograde character, the return to primal unity, occurs again and again as we progress through the numbers. Von Franz sees this as a key to the connection between number and time, and we shall return to this essential point. *Four* is the model of wholeness and completion. The examples from physics suggesting totality or completeness are most impressive: the four dimensions of spacetime, the four basic forces underlying all interactions (gravitation, electromagnetism, and the strong and weak nuclear forces, which physicists of the 1970s and 1980s are striving to merge into one unified force), the four components needed for a complete quantum relativistic description of the electron, and from the primitive sciences, the four basic elements, earth, air, fire, and water. At the same time, in the psyche, four is the symbol of unity and completion—of a person's striving for wholeness, enlightenment, and reabsorption into the One. Indeed, as von Franz says, "Jung devoted practically the whole of his life's work to demonstrating the vast psychological significance of the number four. . . ."[3]

In contrast with this approach, modern science has labored mightily, throughout its history, to take the magic out of numbers. Periodic and regular numerical patterns in nature are always assumed to follow logically from some physical law. They are not considered to have any meaning of their own. For example, the periodic recurrence of chemical properties among the natural elements, as represented in Mendeleyev's periodic table, is not assumed to result from any genuinely magic properties of numbers, such as 2, 8, 10, 18, etc., which show up as the atomic numbers of the chemically similar noble or inert gases. Rather, these numbers are shown to result logically from the quantitative laws describing the symmetries of the atom. And even these elaborate symmetric electron patterns are not attributed to any awesome or mysterious properties of the cosmos as Pythagoras might have done. Instead, these symmetries are seen to result inevitably from the confinement of an electron to a small region of space. The electrons are so confined because of the attraction of the positively charged nucleus. These beautiful and unique atomic patterns with their symmetric lobes, petals, rings, and starlike emanations[4] are in turn reflected at the macroscopic level in endless variety in all the marvelous crystalline structures of matter. The snowflake and the diamond alike owe their superb symmetries to these electron patterns.

And yet we must recall that these symmetric electron patterns are not imprinted on any solid substance or ultimate matter. The electron wave is not a picture of an electron, but only a visual representation of the chances of finding it somewhere in space. The patterns are probability distributions, mathematical fictions, castles in air. They exist in our heads as do the numbers with which we label them. Perhaps, as von Franz suggests, these intricate electron patterns are simply representations (dare I say manifestations?) of numbers themselves. Or are number and symmetry mutual and complementary expressions of an ordered elaboration, a luxuriation of primal inarticulate unity? Are number, order, and symmetry incorporated, as it were, into our expanded spacetime metaphor to safeguard us from the chaotic disorder of unity? After all, why *do* harmony and order seem to suggest peace and security? Do the number symmetries implicit in them lull us with their retrograde and rhythmic motions back to One? And who is it that recognizes and apprehends the numbers and symmetries, and who created them?

> What immortal hand or eye
> Could frame thy fearful symmetry?

The idea of a pattern in space transforming itself into a sequence in time, as suggested by the notion of luxuriation above, is for von Franz the basis of the profound connection between number and time. It all harks back to the quantitative cardinal and ordinal character of numbers. The case is put very well by Tobias Dantzig in his perceptive comments on number as both pattern and sequence:

> Correspondence and succession, the two principles which permeate all of mathematics—nay, all realms of exact thought—are woven into the very fabric of our number system.[5]

Correspondence and succession are important, indeed, but not only in "exact thought."

Von Franz finds this dual aspect of number and its relation to time embodied in ancient Chinese divinatory practices. Two arrangements of the *I Ching* oracles are commonly found: the Ho-t'u "Sequence of Earlier Heaven" and the King Wen "Sequence of Later Heaven." She describes the first:

> The Ho-t'u numerical mandala has no cyclical or linear forms of movement, but rather comprises a kind of ordering of powers or directions that stands as a whole, in a timeless equilibrium. It forms the static image of a greatly intensified inner dynamism; to use a simile, it is like a dragonfly stationary in one single spot in the air, yet held in continuous inner motion through the innumerable vibrations of its wings. . . . The world-model of the Ho-t'u thus forms *the primal image of a relatively timeless state of universal orderedness.*[6]

The idea of an ordered pattern or mandala of pent-up energy and motion held in a state of potential dynamism also characterizes Jung's description of the imaginative archetypes. Jung's profound symbols (or beings) of meaning and order overlap, interpenetrating

Dynamic Dragonfly

and transforming one another into a simultaneous pattern of acausal order, a *spacelike* territory of eternal meaning and correspondence.

The fluttering dragonfly energy of the Earlier Heaven vibrates and occasionally overflows into the Later Heaven, where all is sequence and flight, the temporal causal zone. These erratic outpourings from the Earlier to the Later Heaven appear to us as synchronous events in which one of the eternal and all-pervasive correspondence is revealed to us in a sudden flash. The correspondences of the Ho-t'u seem to us coincidental and sporadic, for the Earlier and Later Heavens are not commensurable. In the King Wen realm of sequential time and causality, simultaneous correspondences and identities are not explicit or apparent. Divinatory practices attempt to open a window through which we may catch a glimpse of the timeless zone. In consulting the *I Ching*, the manipulation of the yarrow stalks provides the opening through which the simultaneous, acausal wisdom of the primal realm may be channeled and transformed into sequential, causal terms to shed light on the course of human events. The astrologer, too, translates the star patterns of simultaneous correspondences into life-historical terms.

Numbers display the same duality through their properties of correspondence and succession; von Franz demonstrates the overlapping correspondences through the retrograde character of number symbolism. *Three* is the synthesis of one and two and therefore represents unity. *Four* is wholeness and completion and thus a new oneness. *Eight* is a double quaternity, a new union of unity and duality. *Ten* stands for a human, a new synthesis, and is also the sum of the first four cardinal numbers. *Eleven*, the number of the Tao, combines five, the center of the first five odd numbers, with six, the center of the first five even numbers, to form a union of oddness (maleness) with evenness (femaleness).

This retrograde, rhythmic, repetitive character of numbers is also apparent in the periodic structure of the elements and in the recurring symmetries of atomic electron states, not to mention octaves in music where higher frequencies or pitches are to us recurrences of lower notes. In numbers, we see the sequence built into the correspondences and vice versa. As we count, we pass through the symbols and in each one we see the ordered reflection of the others. The two concepts are really inseparable. Our perception of them as distinct and incommensurate is tied up with the articulating quality of our cardinal metaphors. Space and number articulate and order experience for us. We see an analogous dual character in time, as well. Acausal, simultaneous, synchronistic time, in which all meaning overlaps, is like the correspondence aspect of number. Causal time, on the other hand, mimics the linear sequence of numbers, with all value strung out in a line, as it were.

Imagine for a moment, a long number axis stretching though empty space. A scintillating point of light moves along the axis, like the spark of a burning fuse, its intensity and color changing as it moves through each number on the line. Flashes of light recur in a periodic and retrograde pattern, but each spark has a unique quality of its own. Viewing the fuse along its length, we see an unfolding sequence of colored flashes. But viewed end on, so that the whole line is telescoped into a point, and eliminating the lapse of time, by allowing all the light images to overlap in a kind of time exposure, we see only a single blinding white vision, with all the many colors and patterns fused into one. That whiteness bears no trace of the flashes of individual numbers. They can only be seen sequentially from the sidelines, as it were. But they are all there nonetheless, in the unifying whiteness. Is number a blinding synthesis or an extended sequence? Or both?

It's intriguing how these time-number images seem to evolve and reveal themselves. Sequential time appears to become wound up and convoluted in its acausal form, which then unfolds and stretches out like an uncoiling snake. Powering the winding and unwinding is the dynamo which von Franz pictures as a fluttering dragonfly suspended in air. Earlier, in chapter 4, we saw another potent image of a primal fluttering rhythm. In the Hindu creation myth, we recognized an incipient breathing as the first step in cosmic manifestation. This pulse of life was seen to be analogous to the role of rhythm in music. Within the mystery of cosmic manifestation was buried the question of how a point can become a line segment: how does zero-dimensional being become one-dimensional being? How does nothing begin its embryonic breathing rhythm and stretch out into something?

The transformation of a point into a line seems to be an alternative and equally mysterious version of the evolution of unity into all numbers or the emergence of sequence out of synchrony. The fusion of the number metaphor with the other cardinal metaphors is becoming apparent. The plural and periodic character of number is analogous to the rhythmic and harmonic nature of music. The mystery of one becoming two is another reflection of the first cosmic eruption and flutter. The retrograde property illustrates again that all is contained in one. The number line, the time axis, and the rhythm wave are unfolded, elaborated versions of the primal unmanifest form. Each point on a line is a recurring image of the fount-point; likewise, each number symbolically contains all the others. Every point, every number, is an image of eternity, "a world in a grain of sand," to use William Blake's words.

At the start of it all is the primal rhythmic energy that leads to that first pulse, the first duality. Which comes first, time-number or energy-rhythm? Von Franz points out that the ancient Greek words for number (*arithmos*) and rhythm (*rhythmos*) have a common origin in the verb *to flow* (ῥεῖν). What precious insights are hidden in the folds of language! But the issue is not precedence; it is representation. Number, time, music, rhythm—all are metaphors of the primal manifestation, of the transformation of *potentia* into *actus*.

The close association of time, energy, and creation in human consciousness is well illustrated by the ancient myths of both East and West. Some primal energy is usually connected with the act of world creation. Plato, in the *Timaeus*, speaks of a basic desire or demiurge which animates the prime mover or *is* the prime mover. According to Jung, energy is the heart of the connection between time and number. He said, "all emotional, and therefore energy-laden, psychic processes evince a striking tendency to become rhythmical."[7] Psychic and mythic events reflect each other. Energy, desire, and sex lie at the heart of creation, of the unfolding of the acausal into the temporal. In the Hindu trinity of Brahma, Vishnu, Shiva, it is Shiva, the destroyer, the agent of world dissolution, who incorporates energy-time-creation metaphors most clearly. For it is the female side of Shiva, his active, power principle, that brings to life his desire and action.

In *Hindu Polytheism*, the author, Alain Daniélou, says,

> The power of Shiva is envisaged under three main aspects: a creative, all-pervading active aspect called Energy (Shakti), a permanent, peaceful, all-pervading, spatial aspect named Parvati, the Daughter of the Mountain (i.e. Ether personified), a destructive, all-pervading time aspect known as the Power-of-Time (Kali).[8]

Shiva's unity is at the same time a dynamic equilibrium between an energy principle, an acausal spatial principle (ether personified, indeed) and a temporal principle. Quoting, in part, from other sources, Daniélou goes on to illustrate the profound intermingling of desire, energy, creation, time, and destruction:

> "The lord-of-sleep (Shiva), the phallus clasped by the womb that is his energy, gives forth the seed of the spatial universe. When conceived as a personified entity, the lord of sleep appears inactive while his energy seems alive. As the instrument of Shiva's procreating power, this energy is the Power-of-Lust. She then appears to be the very opposite of the power of destruction that is Kali, the Power of Time. When Energy, which is also the power-to-think, unites with the lord of sleep, this leads to a state of agitation, of unrest, from which creation springs forth. When she is aloof from him, this leads to the state of sleep, of equalization, in which the world dissolves." When Shakti clasps Shiva, the universe is shaken. She is the all-pervading power of lust, of enjoyment, and also the power of liberation, for liberation from Nature's bonds is not a neutral state but an active fight . . .
>
> It is under her fierce aspect as the Power-of-Time, the power of disintegration closely connected to the power of liberation, that the consort of Shiva is mainly worshipped. She is then shown under a fearful form. She is a fierce-looking goddess, fond of intoxicants, of lust, of bloody sacrifices. Cruel and orgiastic rituals are performed in her honor by the followers of the Tantra cult.[9]

I need hardly point out the associations of energy with thought and of liberation with dissolution and time. The first thought and the first rhythmic energy are different aspects of the same act of creation. We glimpse again in Hindu mythology a view of the physical and mental worlds as one—thought, language, creation, matter. And liberation is entwined with destruction and decay.

For it is only in time that idols may be created and then destroyed. It is energy and desire that motivate all creation, that sustain the world. For the other hidden aspect of Shakti, the energy principle, is Maya[10], i.e., magic or illusion.

Music to My Ears

In chapter 4, we spoke of the cumulative aspect of time and of its unfolding quality as embodied in astrology, acceleration, ritual, and especially music. Von Franz's elaboration of the association and succession properties of number is reminiscent of the earlier discussion of the vertical-harmonic and horizontal-melodic structure of music. Let's pursue the connection between number and music a little further.

Like the number-time metaphors, music may be thought of in two ways: either as an unfolding in a temporal sequence (a piece of music as heard), or as a total pattern all at once (a piece of music as recalled in memory or as written in a score). Benjamin Whorf's idea of cumulative organic time and von Franz's concept of archetypal retrograde numbers are closely related. Both notions are manifest in music with its vertical/horizontal form on the one hand, and its impact as a total experience on the other. Indeed, all the metaphors for number, time, space, and matter become inseparable as we think about them. It is nearly impossible to talk about one without referring to the others. They echo each other like the many reverberations of a call in the mountains. The richocheting sounds appear to come from all sides, but we know they have a single source.

There is another facet of music that can shed further light on the metaphor of number. It is the notion of *interval*, or the difference in pitch between two musical tones. Interval is fundamental to both harmony and melody. A *harmonic* interval refers to two tones sounded simultaneously (vertically), whereas a *melodic* interval applies to the successive (horizontal) sounding of two tones.

Now what is it that determines melody and harmony? Why do we find certain combinations of notes more melodic or harmonious than others? We know that the laws of harmony were first discovered by the Pythagoreans in ancient Greece. Pythagoras, as we have seen, was deeply concerned with numbers. They seemed to him the basis of all form and of geometrical form in particular. He believed that numbers were the foundation of the cosmos. According to G. F. Parker,[11] number for Pythagoras was the eternal unchanging concept on which a stable and secure world could be built. (Always this fear of falling!)

> If then for Pythagoras the basic form, the ultimate truth of the component parts of the universe, was Number, his next logical step was to investigate by what laws the component parts were joined together to make a coherent, working whole. How was universal equilibrium maintained—in so far as it was maintained?[12]

Pythagoras's search for the order and symmetry of the cosmos led him to the laws of musical harmony. He found that the pitch of a note was determined by the frequency of vibration of its sound tone. The more rapidly the string of a violin or the air in an organ pipe is made to vibrate, the higher will be the pitch of the note emitted by these instruments. *Harmonious* combinations of different notes result when their frequencies are in certain numerical ratios. Pythagoras discovered that the Greek lyre was in tune when its four strings vibrated at frequencies in the ratio 6:8:9:12. Otherwise, it was out of tune. Thus only certain pitch intervals or frequency ratios produce harmonious combinations of sound. This idea has been at the heart of all musical composition throughout history.[13]

Pythagoras had his answer: Numbers were the building blocks, and the laws of harmony determined how they were put together into an ordered universe. Certain numerical ratios are harmonious, ordered, symmetric, and these ratios are locked into the structure of the cosmos. But wherein does this quality of harmony lie? Pythagoras saw it as the mean between two extremes, and this pervasive theme has been reflected in all forms of art. For instance, the harmonious pitch combinations of the strings of the Greek lyre are explained by noting that 8 and 9 are respectively the arithmetic and harmonic means of the extremes 6 and 12. But even this explanation begs the question. Why is the mean preferable to the extreme or, for that matter, to any of the other values between or beyond the extremes? Why indeed? Because, simply, *we* say so. It is the human ear and mind that is the ultimate judge of harmony in music —music of the earth and music of the spheres. The remarkable Pythagorean order of the universe has its basis in human aesthetics. When we look out there we see, not space, but our own minds. Where else can Number reside?

Some people argue that the sense of musical harmony may be derived from the laws of acoustics and is therefore basically objective. Let's consider this argument. It is true that the intervals and pitches selected to determine the notes of the Western musical scale—do, re, mi, fa, sol, la, ti, do—can be found among the series of natural overtones of a vibrating string. Suppose, for example, we set a string vibrating at a fundamental resonance frequency of 264 hertz or cycles per second. This is the standard frequency for middle C on the piano. We find not only that the string vibrates at 264 hertz, but also that it produces overtones whose frequencies are integral multiples of 264 hertz: 528 (264 x 2), 792 (264 x 3), 1056 (264 x 4), 1320 (264 x 5), and so on. The musical relationship between the successive frequencies in this overtone series are the standard musical pitch intervals: the octave (528/264 = 2/1), the fifth (792/528 = 3/2), the fourth (1056/792 = 4/3), the third (1320/1056 = 5/4), and so on. Now 528 hertz is the frequency of high C or C′, 792 hertz is the frequency of G′, 1056 hertz is C″, 1320 hertz is E″, and so on. In this way, all the notes of the musical scale, together with their sharps and flats, can be generated.

Thus it might seem that there is a physical basis for the laws of harmony—that the ear selects as harmonious only those pitch intervals between the natural resonant overtone frequencies of a vibrating body. But, in fact, since the time of Bach, musical instruments have not been tuned in accord with the natural overtone series, the so-called just temperaments. Today's instruments, for the most part, are tuned in accord with the more convenient equal temperaments, which differ from the resonant overtone frequencies by enough to sound discordant to a sensitive ear. Furthermore, if we extend the overtone series to ever-higher frequencies (1584, 1848, 2112, 2376, 2640, 2904, 3168, etc.), we can generate quarter tones, eighth tones, and so on, and a series of decreasing musical intervals, many of which we consider inharmonious. In many non-European cultures, different musical intervals are considered harmonious which sound discordant to us. Even in European music, the standards of discord and harmony have changed considerably in the course of history. Every new wave in music, from classical to jazz, has been met with cries of "discordant!" which gradually die away as the unfamiliar harmonies become acceptable and ultimately beautiful.

But the essential point is that music exists in the ear and in the mind, and not in vibrating bodies or in the sound waves they emit. The true nature of a musical experience cannot be described as a sequence of objective pitches, any more than the true quality of a Rembrandt painting can be described as a two-dimensional spatial distribution of colors and light intensities. A listener will often hear a change in pitch when only the intensity or loudness of a note is changed without any corresponding change in frequency. Pitch also depends on the sound environment, i.e., on the pitch and tonality of surrounding instruments. This does not mean that the listener is wrong who hears a change in pitch when there is no change in frequency. Pitch is part of musical experience, and not of acoustics. The listener, especially the experienced musician with a well-trained ear, is the final arbiter of musical quality and meaning. Music cannot exist or be explained apart from a participating listener, or without reference to human experience. Harmony, discord, musical quality, and taste are essentially matters of human choice and value.

In the simple, but profound, human sense that selects certain musical intervals as harmonious—that chooses certain numbers or times or regions and qualifies them with meaning and value—that creates and apprehends symmetry, order, and beauty—in that sense is concealed the dividing, extending, articulating, evolving, creating, destroying power that sustains all our icons. Music does all of this, not as idolatry, but as art.

A Grain of Sand

As a final example of the great power of number as metaphor, I shall survey some of the fabulous, visionary mathematics of Georg Cantor. To my mind, this example illustrates the complete breakdown of the purely quantitative approach to number and space, and shows that the metaphors are inseparable from the measures. Numerical quantity cannot be divorced from meaning. In Cantor's incredible analysis of infinity we clearly see the breakdown.

Cantor stands at a turning point in the history of mathematics, for his researches into the infinite and his invention of set theory heralded for mathematics, "the general collapse of those principles which the prescient seers of the nineteenth century, forseeing everything but the grand débâcle, believed to be fundamentally sound in all things from physical science to democratic government."[14] What an iconclast Cantor was, for, intentionally or not, it was his role to discern such basic faults in the foundations of mathematical logic that the edifice remains unbuilt to this day. Cantor's work on infinite or transfinite numbers was seriously and vigorously challenged once it was recognized how profoundly his paradoxes of the infinite threatened the entire framework of mathematics. The resulting battle of wits has reverberated throughout all of mathematics.

The problem of infinity has plagued mathematics almost from its origin. Not only do the whole integers succeed each other without end, but also, between any two of them, there seems to be an indefinite number of fractions or rational numbers. Zeno's paradoxes and other logical problems of infinity returned to haunt mathematicians after calculus was developed by Newton, Leibniz, and their successors. Calculus clearly worked with amazing success, and yet it was based on notions of infinitesimal and infinite quantities, whose definitions were largely intuitive. This, for the queen of sciences, was an intolerable situation. In the course of the eighteenth and nineteenth centuries, these problems were resolved, and mathematics was once again placed on a rational foundation, through the introduction of the *limit* concept.

In terms of the limit, rigorous definitions may be given for the continuity of numbers (viewed as points on a line), infinite series (never-ending sequences of quantities or mathematical expressions), and for the derivative and integral in calculus. The power of the limit concept, so its supporters believed, lay in its ability to generate such things as derivatives and continuous functions, through a well-defined series of performable steps—a recipe, as it were. It amounts to a prescribed construction or an operational definition. Instead of trying to pursue and locate an infinitely elusive culprit, the limit concept enables us to set a trap around him and settle for the fact that he's somewhere inside the trap. Since the trap can be made arbitrarily small, although always finite, we may get as close to the little devil as we choose. And we pull it all off without ever entangling ourselves in the infinite.

For example, to find the limiting value of the series 1, $^1/_2$, $^1/_4$, $^1/_8$, $^1/_{16}$, etc., we don't chase after this endless series of fractions. Instead, we simply look for a number that is within an arbitrarily small range of all the members of the series beyond some chosen one. The number 0.1 cannot be the limit because, although it is within 0.1 of all the members beyond $^1/_8$, it is not within 0.01 of any of them. In other words, 0.1 is not as close as we might please to the members of the series. Similarly for 0.01, 0.001, 0.0001, and so on. Clearly, the only number we can catch in an arbitrarily small trap is 0, and 0 is therefore the limit. We never actually reach 0; we guessed it from the way things were going. But we know we can come as close to the limit as we choose through a *finite* number of steps.

Now the whole limit idea is something of a fake, of course, because there's a difference between owning a fur coat and being told that the hunter is arbitrarily close to trapping the mink. Nevertheless, the limit concept and the painstaking labors of over a hundred years had convinced the mathematicians of the late nineteenth century that they were within sight of the long-sought-after goal of *axiomatizing* mathematics a la Euclid, i.e., of logically deriving all the concepts, theorems, and methods of mathematics from a small number of assumed axioms. At least, so they thought until Cantor came along, and the struggle against him was long and bitter. (It may even have contributed to his final mental breakdown.) The validity of Cantor's methods of dealing with infinite sets was seriously questioned, and the basic paradoxes were discovered even in their definition. Fundamentally, Cantor's theory was nonconstructive: it assumed the existence of things that could not be constructed through a finite number of steps. But mathematicians were unwilling to reject Cantor's brilliant work because it was not constructive or axiomatic any more than they would earlier have considered rejecting calculus because it could not be rationally justified.

Furthermore, the axiomatic-constructive program seems doomed to failure, at least in its original terms. The brilliant mathematician Kurt Gödel published in the 1930s a revolutionary series of papers which demonstrated that any branch of mathematics, sufficiently complex to be of "interest" to mathematicians (this includes arithmetic, geometry, and calculus among others) will be either inconsistent or incomplete.[15] This is a devastating result (which was obtained, incidentally, according to strictly constructive and unimpeachable procedures). It says, for instance, that if you try to derive the laws of arithmetic in a logical fashion from any set of postulates, simpler than arithmetic itself, either you will arrive at contradictory results (inconsistency) or there will be results that you know to be true, but which you cannot derive from the postulates (incompleteness). And, as if this blow weren't sufficiently crushing, Gödel goes on to show that these inconsistent and unprovable results are neither trivial nor few, but instead, are both profound and infinite in number. This may well mean that mathematics can never be put on a firm rigorous foundation, let alone physics or anything else. Therefore, we may safely turn to the marvels of Cantor without too much concern for their ultimate logical foundation.

In his attempt to define and come to grips with numbers, Cantor invented the concept of a *set* and the notion of one-to-one correspondence between the members of different sets that we have already used in our discussion of measurement in chapter 1. Cantor was getting at the essence of cardinal numbers and their ability to describe amount or size. Two sets are said to have the same *cardinality* when their elements can be placed in a unique one-to-one correspondence with each other. If I have a set consisting of a cat and a kite, I can compare it with my standard sets of elephants. By placing the cat on Howard's back and asking Dorothea to hold the kite in her trunk, I have neither elephants, cats, nor kites unaccounted for and so I know that the two sets have the same cardinality, viz., 2.

All of this begins to take on new interest when we realize that

Cantor's method enables us to deal with a set having an infinite number of elements. To count the elements of an infinite set would take forever; that's impractical. But if we can guarantee a one-to-one correspondence between the elements of two infinite sets, we know that they have the same cardinality. If one of them is a standard set of known cardinality, we have determined what the cardinality of the other is. As a standard set with the cardinality of infinity, Cantor chose the set of positive integers: 1, 2, 3, 4, and so on. This choice is natural enough for the integers are a prototype for many other infinite series and sets. Cantor needed a name for this new infinite cardinality. In order not to confuse it with any of the names for finite sets (like 2 or 1,278,033) and so as to distinguish it from all other common symbols of mathematics, he selected the first letter of the Hebrew alphabet, aleph. It is written $\aleph_0$ and pronounced aleph-null. The reason for the subscript zero is to distinguish $\aleph_0$ from all the other infinite cardinalities Cantor discovered, which he labeled $\aleph_1$, $\aleph_2$, and so on. To his amazement, the first thing Cantor found was that many infinite sets have the cardinality $\aleph_0$. This might not seem too surprising for we should not be disturbed by two finite sets having the same number of elements. What startles us are the particular infinite sets Cantor found with the same cardinality as the integers. We intuitively assume, for example, that there must be fewer even numbers than there are whole numbers; we should expect there to be half as many, but in fact, there are the *same* number. This is easy to prove: every even number may be associated with an integer in a unique way, leaving no members of either set out of the correspondence:

$$\begin{array}{cccccccccccc} 1 & 2 & 3 & 4 & 5 & 6 & 7 & 8 & 9 & 10 & \ldots \\ \updownarrow & \updownarrow & \updownarrow & \updownarrow & \updownarrow & \updownarrow & \updownarrow & \updownarrow & \updownarrow & \updownarrow & \\ 2 & 4 & 6 & 8 & 10 & 12 & 14 & 16 & 18 & 20 & \ldots \end{array}$$

Since this is a unique one-to-one correspondence between two different sets, we must assume as a matter of definition that they have the same cardinality, just as we should do in the case of finite sets. It may seem to violate intuition, but we are dealing here with infinite magnitudes and we do not know how far our intuition is to be trusted.

The same argument can be made for the odd numbers. We need only display the correspondence:

1	2	3	4	5	6	7	8	9	10	. . .
↕	↕	↕	↕	↕	↕	↕	↕	↕	↕	
1	3	5	7	9	11	13	15	17	19	. . .

So the odd numbers also have cardinality $\aleph_0$. This, however, leads us to an astounding result. The set of whole numbers consists precisely of the set of even numbers plus the set of odd numbers. Symbolically:

$$(2, 4, 6, 8, \ldots) + (1, 3, 5, 7, \ldots) = (1, 2, 3, 4, 5, 6, \ldots)$$

Now one of the main reasons for introducing sets in the first place is that they enable us to study the arithmetic properties of numbers. This was the key to their use in the new math programs in elementary schools. By combining a set of two ducks with a set of three pigs, children easily see that the compound set has five elements, and so they learn the laws of addition. When two sets are combined, their cardinalities add. When we write $2 + 3 = 5$, we are illustrating a special case of a very general property of sets. But this argument forces us to Cantor's conclusion:

$$\aleph_0 + \aleph_0 = \aleph_0 \quad \text{or} \quad 2 \times \aleph_0 = \aleph_0$$

There's no misprint here. The sum is *not* $2\aleph_0$ but $\aleph_0$. The two sets combined each have a cardinality of $\aleph_0$, which we know by comparing each of them with a standard set of integers. But the two sets *combine* to form the standard set of integers, so that the resulting set also has the cardinality $\aleph_0$ by *definition*. This, then, is the first paradoxical law of transfinite arithmetic, *the whole is no greater than its parts*.

This remarkable result tells us that when we are dealing with an infinite set, some of its component sets, or *subsets*, have the same cardinality as does the original set. And this is true not just for a small number of specially selected subsets, but for an *infinite* number of subsets. Many other examples can be given. The sets (10, 20, 30, . . .) and (1,000,000, 2,000,000, 3,000,000, . . .) and many like them can clearly be placed in one-to-one correspondence with the integers, and so they all have cardinality $\aleph_0$.

Or take the set (2, 4, 8, 16, 32, 64, 128, 256, . . .) whose elements can be rewritten as powers of 2, (2^1, 2^2, 2^3, 2^4, 2^5, 2^6, 2^7, 2^8, . . .). In this latter form the set has a manifest one-to-one correspondence with the set of integers (the powers are the integers), so it too has cardinality $\aleph_0$. This is true of so many subsets of the integer set, which in turn combine to form the integer set, that $\aleph_0$ multiplied by *any number* is still $\aleph_0$. In fact, since it is true for an *infinite* number of subsets of the integer set, it follows that

$$\aleph_0 \times \aleph_0 = \aleph_0 \quad \text{or} \quad \aleph_0^2 = \aleph_0$$

That is, $\aleph_0$ times itself is $\aleph_0$.

By now it might seem that all infinite sets have the same cardinality, $\aleph_0$, and that puts an end to the matter. We shall see that this is not the case. But first we shall consider an example that shows just how pervasive $\aleph_0$ is among infinite sets. So far we have looked only at subsets of the integers. What if we consider sets obviously "larger" than the integer set? A set like ($^1/_2$, 1, 1$^1/_2$, 2, 2$^1/_2$, 3, 3$^1/_2$, 4, 4$^1/_2$, . . .) will still have $\aleph_0$ members since we can simply label each element with an integer. (Doubling this series, in fact, produces the integers). But let's consider a far more overwhelming case, the set of all fractions.

It might seem preposterous to suggest that there are no more fractions than integers, but that's exactly what Cantor proved. And the result applies not only to fractions less than 1, like $^1/_2$, $^2/_9$, $^{3256}/_{4301}$, or $^{9,999,999,999}/_{10,000,000,000}$, but also to all mixed fractions greater than 1, such as $^4/_3$, $^{27}/_{13}$, $^{23,089}/_7$, or 38. This last example, 38, is actually the fraction $^{38}/_1$ or $^{76}/_2$ or $^{114}/_3$. In other words, all the integers are already included in the set of all possible fractions. This set is usually called the set of *rational* numbers because its members are all the possible ratios of the integers. Every conceivable ratio of two integers is a rational number. And this fact enabled Cantor to establish the one-to-one correspondence between the rational numbers and the integers. Cantor's proof is illustrated in Figure 6-1, in which *all* the rational numbers are displayed in an array, ordered arbitrarily according to the sum of their numerators and denominators. They are all there because every possible combination of numerator and

Sum of Numerator and Denominator N+D	Corresponding Rational Numbers N/D							
2	$^1/_1$							
3	$^1/_2$	$^2/_1$						
4	$^1/_3$	$^2/_2$	$^3/_1$					
5	$^1/_4$	$^2/_3$	$^3/_2$	$^4/_1$				
6	$^1/_5$	$^2/_4$	$^3/_3$	$^4/_2$	$^5/_1$			
7	$^1/_6$	$^2/_5$	$^3/_4$	$^4/_3$	$^5/_2$	$^6/_1$		
8	$^1/_7$	$^2/_6$	$^3/_5$	$^4/_4$	$^5/_3$	$^6/_2$	$^7/_1$	
.	.	.	.	.	.	.	.	
.	.	.	.	.	.	.	.	...

Figure 6-1.
Denumerable array of rational numbers

denominator would ultimately appear on this diagram. By moving vertically downward through this diagram and listing each horizontal line of fractions as we come to it, we would eventually produce a list of *all* the fractions (or rational numbers) in one continuous line:

$$^1/_1,\ ^1/_2,\ ^2/_1,\ ^1/_3,\ ^2/_2,\ ^3/_1,\ ^1/_4,\ ^2/_3,\ ^3/_2,\ ^4/_1,\ ^1/_5,$$
$$^2/_4,\ ^3/_3,\ ^4/_2,\ ^5/_1,\ ^1/_6,\ ^2/_5,\ ^3/_4,\ ^4/_3,\ ^5/_2,\ ^6/_1,\ ^1/_7,\ \ldots$$

We could then label each of these fractions in succession with an integer. This labeling process provides the necessary one-to-one correspondence with the integers to prove that the cardinality of the rational numbers is none other than $\aleph_0$. Not only do infinitely many sets contained among the integers have cardinality $\aleph_0$, but so do infinitely many sets containing the integers. The laws of transfinite arithmetic are trivial: $\aleph_0$ plus, minus, times, or divided by anything else is still $\aleph_0$. No matter how you slice it, it's always the same.

Now we come to a new level of marvels in Cantor's uncanny analysis of the infinite. It may be almost impossible to believe by now that there is any infinite set of numbers not having $\aleph_0$ for its cardinality. But Cantor exhibits one for us. The key to this is the idea of *denumerability* or countability. All the infinite sets of numbers we have looked at so far have been denumerable. In other words, they have been sets of discrete or distinct elements, which can be separated from each other and counted, at least in principle. You can count the ducks on a pond, the leaves on a tree, even the atoms in a rock, because each element in these sets (a duck, a leaf, or an atom) is separate from every other element. These three are examples of finite sets, large perhaps, but still finite. But the same character of denumerability holds for many infinite sets. In fact, our prototype set, the integers, is essentially an abstraction for the very notion of counting or denumeration. Of course, we cannot actually count the members of an infinite set, and that's why Cantor invented set theory and the notions of cardinality and one-to-one correspondence. But it is possible in principle with all the sets we have examined so far. So, to exhibit a set of cardinality greater than $\aleph_0$ we must answer the question: does there exist an infinite set that is *nondenumerable*? Is there a set with infinitely many elements but which can *not* be put into one-to-one correspondence with the integers?

The answer, according to Cantor, is the set of all *real numbers*. The real numbers (they are called real to distinguish them from the so-called imaginary and complex numbers, which need not concern us) include, in addition to the integers and rational numbers that we've already discussed, a group of numbers called *irrational*. These numbers aren't irrational by virtue of being mentally defective (although their inventors were considered so by some people), but simply because they are not rational, i.e., they cannot be represented by the ratio of two integers. The square root of 2 is an example of an irrational number. The fraction $^{707}/_{500}$ is a reasonable approximation to the square root of 2; when squared it gives 1.999396, which isn't bad, but it isn't 2. There is no fraction that when squared will give exactly 2, although one may get as close to it as one pleases.

Not all square roots are irrational. The perfect squares like 4, 9, 16, 25, 36, etc., all have rational square roots. But most numbers have irrational square roots. And then there are the cube roots, the fourth roots, the fifth roots, and so on ad infinitum. In addition, the irrationals include many numbers that are not even roots of rational numbers. For example, there is e, the base of the natural logarithms, and π, the ratio[16] of the circumference of a circle to its diameter. These particular irrational numbers are called *transcendental*. There are apparently a good many irrational numbers. The problem is to figure out how many. After all, there are a lot of fractions and integers, too. Does adding the irrationals to the rationals to form the real numbers produce a set with some new cardinality, greater than $\aleph_0$? The answer is yes, and Cantor proved it.

His proof is ingenious, and we shall sketch it.[17] The proof is by contradiction. Cantor assumes that it is possible to make a listing of all the real numbers and then proceeds to show that this assumption is false. He exhibits a symbolic list of numbers in decimal notation which is, by assumption, a *complete* list of the real numbers. This assumption is valid because it is always possible to represent *any* number, rational or irrational, as a decimal. He then produces one number that is not in the list and which can never be contained in any such list of decimal numbers. Since Cantor has thus demonstrated that every possible list of decimals has at least one number missing from it, the original assumption that the list was complete must be false. Furthermore, the method by which Cantor constructed the one missing number can be used to construct infinitely many more.

Now why does this prove that the real numbers are nondenumerable? Because Cantor has shown that a *listing* of them is impossible. And a list is precisely what is implied by denumerability. A list is a series of distinct entries and is therefore countable. We may always number each separate entry in a list and thus form a one-to-one correspondence between any list and the set of integers. So any infinite list is denumerable and must have cardinality $\aleph_0$. But Cantor has shown that the real numbers can *not* be listed and so are nondenumerable. And since there must be more real numbers than there are integers, because infinitely many of them are left out of any denumerable list, then the cardinality of the real numbers is greater the $\aleph_0$. Cantor succeeded in exhibiting an infinite set, the real numbers, that *cannot* be put into one-to-one correspondence with the integers, and which is obviously larger than the set of integers. The cardinality of the real numbers Cantor dubbed C for continuum, and he called it the *cardinality of the continuum*.

C is the cardinality of all the real numbers. They form a set of continuous elements like the points on a line. Neither the real numbers nor the points on a line are denumerable. Both sets are continuous, as opposed to discrete, and are nondenumerable. In fact, the real numbers and the points on a line may be placed in a perfect one-to-one correspondence, demonstrating their fundamental equivalence as sets of equal cardinality, C. This is why an infinitely long axis may be used as an exact geometric representation of the real number system.

Cantor proved that C is greater than $\aleph_0$. But he was not able to prove that C is equal to $\aleph_1$ (aleph-one), the first in a long line of infinite "magnitudes" above $\aleph_0$. It remains unproven to this day. Are there infinite sets of cardinality intermediate between $\aleph_0$ and C? Is C really the first after $\aleph_0$? Are the cardinalities even denumerable? These unsettled questions continue to open new mathematical vistas.[18] But for now, we'll settle for a peek at the truly fantastic realm of C.

As with $\aleph_0$, we will find many different sets with cardinality C, but here the correspondences are even more astounding. First we find that a long line has no more points than a short one. The standard of comparison is, as usual, a one-to-one correspondence. In Figure 6-2, we see two arbitrary straight line segments AB and CD. They are arranged so that any ray projected from the point O will pass through one and only one point of each line. Thus any point in AB such as P can be uniquely associated with only one point in CD, such as Q. And this is true for every point in both lines. Thus the diagram illustrates a perfect one-to-one correspondence between the points in the two lines. Considered as sets, or point sets as they are called, AB and CD have the same cardinality. Clearly, this can be proved for any two line segments, and therefore all line segments have the same "number" of points regardless of their length.

Furthermore, the cardinality of such point sets (those that are

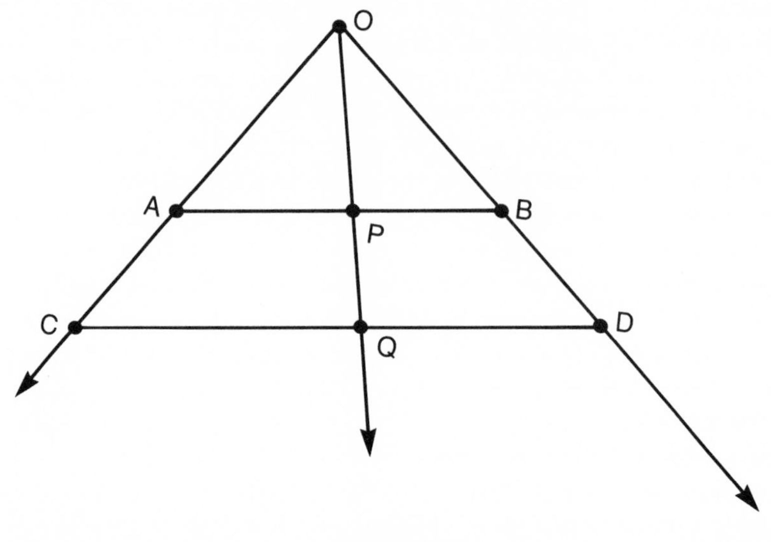

Figure 6-2

one-dimensional and of finite length) is C. But C is the cardinality of the real numbers or of an *infinitely* long line. Can it be that an infinite line has no more points in it than a finite one? The answer is yes. It is not hard to understand why: Doubling a line segment does not change its cardinality, nor does tripling it, quadrupling it, and so on. As we've just seen, the cardinality is independent of the length. This means that multiplying C by any number does not change its value; it remains C. We had the same law of multiplication for $\aleph_0$, so this result should not be surprising. But we also discovered that $\aleph_0$ multiplied by itself is still $\aleph_0$. If we combine $\aleph_0$ sets, each having $\aleph_0$ elements, the total set still contains $\aleph_0$ elements. The same is true for C: $\aleph_0 \times C = C$. No matter how many sets of cardinality C we throw into the pot, we still end up with a set of cardinality C. Even if we add up a denumerably infinite number ($\aleph_0$) of sets of cardinality C, we get a set of cardinality C. An infinitely long line contains no more points than any finite segment of it. After all, the segments are denumerable, and so there are only $\aleph_0$ of them. But the points in each segment are nondenumerable; there are C of them. So we are adding up $\aleph_0$ segments, each having C points. And C is so "large" an infinity that even having $\aleph_0$ of them cannot increase it. Of course, this is even true of $\aleph_0$ itself. It's as if all infinities swallow up all numbers and infinities smaller than themselves, and *even* themselves.

But is this last rash statement really true of C? Is C x C = C? Do C sets of cardinality C form a set of cardinality C? The incredible answer, once more, is yes. And Cantor's proof will carry us to the most dizzying heights of infinity. For he demonstrates that even a two-dimensional continuum, like a square, contains C points. There are no more points in a square than there are in any line segment that borders it. Once again, I refer you to Kasner and Newman for the details, but the essence of the proof is simple: Cantor shows that each point in a square uniquely corresponds to only one point on its edge. Thus the square point set has the same cardinality as its edge, C. This proves that C x C = C. For a square may be thought of as consisting of an array of parallel line segments, like a picket fence. Of course, the line segments are infinitely concentrated: there are C of them. But since each of the C line-segment, picket elements contains C points, the number of points in the whole square is C x C. But by Cantor's proof, the number of points in the square is C, and so the uncanny law of multiplication is proven.

It is now a simple matter to generalize Cantor's proof from two dimensions to three, and to show that a cubical point set also has cardinality C. There are no more points in a cube than in one of its square faces or in one of its edges. A volume in three-dimensional space is equivalent, as a set of points, to a one-dimensional line segment. The mysterious boundaries between dimensions seem to evaporate when space is treated as a continuum of points. The points in a cube may be rearranged to form a line segment without losing or gaining a single one. All the myriad points that inform the cube and give it a felt extent in three dimensions may be funneled together, as it were, into a mere thread, having length but no thickness or width. And this is true not just for the finite cube, but for all of infinite space. For space may be built up out of cubes as a line is from segments. In all three-dimensional space, there are only $\aleph_0$ cubes. In other words, all the cubes filling space are denumerably infinite. Since each cube contains C points, so does *all* space ($\aleph_0$ x C = C). And since the shortest of line segments contains as many points as the longest, all the vast regions of infinite space may be coalesced into some tiny line segment, like a hyphen. What metaphor could better suggest the collapse of space into a miniscule unity? Or the reflection of the infinite complexity of all things in each other? But William Blake said it better:

> To see a World in a Grain of Sand
> And a Heaven in a Wild Flower,
> Hold Infinity in the palm of your hand
> And Eternity in an hour.

And yet we are still short of that final fall into the point of no size at all. Can Cantor do it for us? Can Blake? Can anyone?

Fugue in Aleph

Cantor's analysis of the infinite, presented as the series of proofs we have just surveyed, has always been very special to me. I find it more a work of art than of mathematics, if such distinctions can be drawn at all. In the attempt to come to grips with the very essence of counting, the work begins, like a fugue or a set of variations, with the deceptively simple theme of one-to-one correspondence. Working its magic with the harmony and melody of numbers and point sets, the theme is repeated again and again in varied and startling form, revealing new delight and meaning in the infinite. The music rises in a series of crescendos, sweeping us up to a peak with an astonishing vista of the universe in the merest speck. The climax subsides. We return from the heights with a sense of ineffable comprehension. We pass from the infinite to the finite in the final resolution of a chord.

I believe that in such evocative uses of the theories and concepts of mathematics and science lies their greatest human value. Like all art, science stems from the quest for meaning, purpose, and union, and it serves and satisfies us best through its allegories and symbols which portray that great quest and its goals. But this exalted function is constantly frustrated by our literal interpretation of science and its handmaidens. We pay lip service to science-as-metaphor, but refuse seriously to integrate this idea into our everyday practice of science and, above all, into our teaching of it. We occasionally sing the praises and beauties of science, but rarely

To see a World in a Grain of Sand
And a Heaven in a Wild Flower.

in our classrooms. We feel these aesthetic experiences to be too personal and private to be communicable, and too subjective and inappropriate to stand amid the familiar stone colossi of objective science. But these cold stone giants are idols of arrogance. If the scientist does not risk infusing them with human blood and spirit, who will? Who will stem the rising tide of suspicion about scientific idolatry while there is yet time to catch the floodwaters at their source? Who will prevent the deluge and the destruction of our present idols and their replacement by the contrary, and equally inhuman, icons of unreason and madness? We must use the metaphors of science and the physical world to nourish our lives and souls, and not to feed our fears of falling and dying.

Chapter 7.

Death and Transfiguration

We turn now to the psychology of the cardinal metaphors of physics. What are the motives behind them? How do they reflect our psyche? What basic human needs or fears do they express? I was led to these questions by contemplating scientific metaphors in the light of the central thesis of this book, that physical reality is a creation of the mind. I shall now go on to explore some of the psychological implications of this thesis.

In the foregoing chapters, I have tried to show that the quantitativeness of physics does not guarantee an objective physical world and that there is no meaningful boundary—indeed, no real difference—between the subjective and objective approaches to things. In all of this, I have conceived of consciousness as evolving. Reality, a projection of consciousness, must also evolve and is a timely expression of the state of the human mind and spirit.

As we explored space, time, matter, and number we saw that their purely quantitative aspect, as used in physics, is inseparable from our subjective intuitions of them. But we saw as well that these cardinal metaphors deal with our deepest questions and fears about our existence. They express our elemental needs for extension, distinction, identity, stability, endurance, variation, movement, and meaning. Even their quantitative character seems to fortify us with a deceptive measure of mortality and existence. We know we shall not last forever, but we clutch at the assurance of a canonical threescore and ten years of life (or, at the latest prediction from the mortality tables). We fool ourselves and hang on to enough sanity to remain functioning animals through the magic of number and extension. But the world beyond our metaphors and illusions is unitary and chaotic. It is a world filled with the terror of death. It is with that terror that we shall now come to grips.

The Denial of Death

The principal motive behind our scientific metaphors is the avoidance of death. I must confess my strong bias in this matter, if indeed it is not already apparent to the reader. Through my own experiences, recollections, and meditations upon why we do the things we do, I have come to believe in our need to create illusions to escape the fear and terror of death. This powerful drive motivates most human activity. The fear of death underlies most psychological, existential causes, preceding all others. I have been strongly influenced in this line of thought by Sigmund Freud, Brigid Brophy, and Norman O. Brown,[1] and most recently and profoundly by Ernest Becker in *The Denial of Death*.[2] Despite the insight of Karl Marx and others, I do not believe that the fundamental cause of human suffering is purely economic and material. This in no way is meant to absolve us from any moral responsibility to our fellow humans. On the contrary, our illusion-building and creative potentials bind us mutually and morally all the more to each other. But more of this later.

Because of the central role *The Denial of Death* has played in my own thinking, I shall try now to summarize Becker's principal ideas. There is, however, an additional reason for this summary. In spite of my admiration and respect for Becker's synthesis of psychology, religion, and existential philosophy, there is in his book a certain weakness I wish to point out. Ironically, this weakness is also one of its great strengths, and quite relevant to this discussion of metaphor. Becker seems to harbor a certain ambivalence on the matter of a person's *leap into faith*. This was Kierkegaard's term for the soul's agonized but resolute choice, catapulting it from existential despair into religious faith. I feel that the problem of despair-in-freedom vs. faith-in-bondage was not entirely resolved in Becker's own mind, not from any lack of Becker's understanding or subtlety, but rather because of his great integrity. We shall never know the full truth, for Becker died after completing his book.[3] It is as if with his own life and death he reemphasized the awesome choice we each must face alone.

Becker begins with the paradoxical duality of a human's experience as an ethereal self in a material body:

> Man has a symbolic identity that brings him sharply out of nature. He is a symbolic self, a creature with a name, a life history. He is a creator with a mind that soars out to speculate about atoms and infinity, who can place himself imaginatively at a point in space and contemplate bemusedly his own planet. This immense expansion, this dexterity, this ethereality, this self-consciousness gives to man literally the status of a small god in nature, as the Renaissance thinkers knew.
>
> Yet, at the same time, as the Eastern sages also knew, man is a worm and food for worms. This is the paradox: he is out of nature and hopelessly in it; he is dual, up in the stars and yet housed in a heart-pumping, breath-gasping body that once belonged to a fish and still carries the gill-marks to prove it. His body is a material fleshy casing that is alien to him in many ways—the strangest and most repugnant way being that it aches and bleeds and will decay and die.[4]

We are driven by the terrible inner contradiction we feel: each of us is an infinitely expansive symbolic self imprisoned in a decaying and death-bound body. We must struggle and fight heroically to claim our own meaning and value with every symbolic weapon in our arsenal:

> This is what society is and always has been: a symbolic action system, a structure of statuses and roles, customs and rules for behavior, designed to serve as a vehicle for earthly heroism . . .
>
> It doesn't matter whether the cultural hero-system is frankly magical, religious, and primitive or secular, scientific, and civilized. It is still a mythical hero-system in which people serve in order to earn a feeling of primary value, of cosmic specialness, of ultimate usefulness to creation, of unshakable meaning. They earn this feeling by carving out a place in nature, by building an edifice that reflects human value: a temple, a cathedral, a totem pole, a skyscraper, a family that spans three generations. The hope and belief is that the things that man creates in society are of lasting worth and meaning, that they outlive or outshine death and decay, that man and his products count.[5]

There are those today who argue that heroism is bourgeois and obsolete,that we are conditioned to it by a prejudiced and elitist society, and that it has no place in the postrevolutionary world of antiheroes and nonindividualists. This argument puts the cart before the horse. Heroism is not the cause but the symptom of our disease; and that disease is the human condition itself, not just a material condition, but an existential one. We are conditioned to heroism through our dawning consciousness of the human plight. As children grow and develop awareness, it is the inarticulate fear and terror of their animality and the struggle to overcome the limitations and contingencies of their bodies that lead them to imitate adults by repressing the fact of mortality and to validate themselves through heroic acts. People become prejudiced, cruel, inhuman, and hungry for power and wealth because these are conventional compensating mechanisms for our fear of death and annihilation. Of course, society is wrong and even immoral in providing these self-defeating mechanisms, but the real problem is not that the few are unfair to the many (however true that may be), but that life is unfair to us all. It is not heroism that is at fault, but false and idolatrous heroism.

Becker calls all these false heroics and life-styles *vital lies*, which are necessary because our false illusions are used to maintain the equanimity to function in our everyday lives. The raw unmitigated reality of life and death would throw most of us into a paralyzing state of fear and trembling, of absolute chaos without relief. Who wouldn't take on a few illusions and neuroses to avoid that? So we become involved in a "second-hand" struggle, not with despair itself, but rather with our "screen against despair," with the vital lies, with the

> stock market, with sports cars, with atomic missiles, with the success ladder in the corporation or the competition in the university. . . . Even in our passions we are nursery children playing with toys that represent the real world. Even when these toys crash and cost us our lives or our sanity, we are cheated of the consolation that we were in the real world instead of the playpen of our fantasies. . . . It is fateful and ironic how the lie we need in order to live dooms us to a life that is never really ours.[6]

Becker says we are torn by our ironic dual nature, which is the human condition. We feel ourselves to be symbolic, ethereal, timeless beings, and yet we know we are creatures of flesh and matter, subject to decay and death. We struggle heroically against this cruel paradox, creating symbolic systems in society, science, religion, the home, and the mind to reinforce ourselves and give us meaning. These vital lies provide a screen against despair and the awesome horror of a meaningless life and death, but they also prevent us from facing the truth squarely. Our life struggles are secondhand and illusory.

This analysis of human motivation and action in terms of the fundamental and inalienable dilemma of life enables Becker to present a theory of neurosis and mental illness as "failed heroics." The details of Becker's brilliant theory do not directly concern us, but the spirit of his approach to neurosis and psychosis as faulty illusion building and failure of courage is particularly relevant to our discussion of consciousness and the metaphors of science.

Schizophrenia, or Life without Metaphors

I am a little embarrassed to be writing about schizophrenia. I'm not a psychologist; nor do I have any clinical experience. But I feel compelled to deal with this subject since the mind of the psychotic can reveal so much about our own illusions. I shall rely on Becker and R. D. Laing, whose insights, I believe, support the view of "normal" consciousness as idolatrous.

Often in this book, I have described the consciousness characteristic of some ancient and remote society to illustrate a world view alternative to ours. I used, for example, the astrological consciousness of a medieval mind to illustrate some very unfamiliar perceptions of spacetime. Such altered states of consciousness can no longer be looked upon as quaint curiosities. In recent times, they have been experienced and observed quite close to home. The burgeoning literature on drug-induced states by Aldous Huxley, Alan Watts, John Lilly, Charles Tart, Claudio Naranjo, and Andrew Weil, to name a few, forces us to consider the matter seriously. And Gregory Bateson, Thomas Szasz, Becker, and Laing, among others, enable us to see psychotic states not only as relative and socially defined, but even (and this is especially true of Laing) as a natural path toward health and enlightenment.

Becker shows us that neurotics and psychotics are, in a sense, more in touch with reality than the so-called normal person. (These categories are all somewhat arbitrary, but at least they are useful in describing and analyzing the compulsive behavior that we all exhibit to some degree. It's worth remembering, however, that extreme psychotic states lie on a continuum with normal behavior.) The basic nature of psychosis, according to Becker, is that mentally disturbed people are ineffectual in using the defense mechanisms that most people employ. They cannot live the vital lie. They have not succeeded in constructing a screen against despair which they can rely on to maintain their equanimity. The schizophrenic may not even have the secure sense of being grounded in a body which he or she can take for granted and use to manipulate the objects of the world. Becker's description of the schizophrenic is graphic and devastating:

> In his early childhood development he did not develop a secure "seating" in his body: as a result his self is not anchored intimately in his neuroanatomy. He cannot make available to himself the natural organismic expansion that others use to buffer and absorb the fear of life and death. . . . We know today that the cultural sense of space, time, and perception of objects are literally built into the neural structure. As the cultural immortality ideology comes to be grounded in one's muscles and nerves, one lives it naturally, as a secure and confident part of one's daily action. We can say that the schizophrenic is deprived precisely of this neurological-cultural security against death and of programming into life. He relies instead on a hypermagnification of mental processes to try to secure his death transcendence . . .
>
> Schizophrenia takes the risk of evolution to its furthest point in man: the risk of creating an animal who perceives himself, reflects on himself, and comes to understand that his animal body is a menace to himself. When you are not even securely anchored in this body it really becomes a problem. Terror becomes unabsorbable by anything neural, anything fleshy in the spot where you stand; your symbolic awareness floats at maximum intensity all by itself. This is really a cursed animal in evolution, an animal gone astray beyond natural limits. We cannot imagine an animal completely open to experience and to his own anxieties, an animal utterly without programmed neurophysical reactivity to segments of the world. Man alone achieves this terrifying condition which we see in all its purity at the extremes of schizophrenic psychosis.

> In this state each object in the environment presents a massive problem because one has no response within his body that he can marshal to dependably respond to that object.[7]

Apart from the power of such a passage, I have quoted it extensively because I see the schizophrenic state of disembodied existence, in which each object presents a "massive problem," as a model for nonidolatrous experience, for life without illusions.

This terrifying state is not too difficult to imagine or even to experience temporarily. Have you ever, in a dream, experienced the horror of falling, or dying, or of being overtaken by an all-consuming blackness? Have you ever lost, or imagined losing, a limb or the power to control your own body? Can you picture being so handicapped that an ordinary object becomes a "massive problem?" These and countless other common experiences involving such feelings as helplessly falling or floating, dissolving, being trapped or buried alive, losing control, or losing your body altogether, all suggest a state of ethereal awareness with no scaffolding or support of any kind, nothing to "buffer and absorb the fear of life and death . . . in the spot where you stand." This is perhaps the first step toward nonspatiality. If you completely lose your very physicality, where and when do you exist, persist, subsist? Blindness, deafness, anesthesia, paralysis, disembodiment —pure ethereal thought and nightmare images. How close is this to death itself, to the absence of space and time? Here are all the fears and forebodings that we associate with death and oblivion, all the childhood nightmares we have put behind us and never dare contemplate again. Here is the awesomeness of death and of chaotic helpless existence. Here in this spaceless devouring state I can imagine myself as a wretched schizophrenic with no space to give me *lebensraum*; no time to let me move in it; no matter of my own to move and control and with which to negotiate being; no law of inertia to keep me effortlessly self-supporting; and no numbers with which to divide and conquer my environs. I stand before shattering existence, myself mere existence. I am forced to face raw life and death without a shred of my own meaning, without a metaphor to my name. Is it any wonder that I passionately create my own world, that I contrive elaborate fantasies, that I rely on a "hypermagnification of mental processes" to try to secure my death transcendence?

But unlike the schizophrenic, we "normal" types have no need for ephemeral personal fantasies. We exist in space and time, securely localized in solid bodies, supported by the massive earth relentlessly obeying Newton's Universal Law of Gravity (or is it Einstein's inertial geometry?—Well, no matter, someone's law holds us up). What's more, we have numbers with which to count the atoms and the stars, weigh the universe, and measure its age. With such a shared reality, what have we, the sane and normal, to fear? Death? Well, I suppose so. We must all die. But why dwell on it? Why be morbid? Besides, I can't think about it anymore. I've got an appointment to keep, which means that I can occupy myself with manipulating and transporting my body and arranging to locate it in a certain place at a certain time. And afterward there'll be endless combinations of space, time, matter, and number to jockey and occupy myself with. What more could anyone ask?

Yet there is more to glean from this fantastic schizophrenic state. My mind keeps returning to the potent and terrifying image of object as *massive problem*. I sense some great looming presence before me over which I have no control and which I cannot comprehend. It threatens me utterly. It can smash me, devour me. Its very being is ominous and mysterious. I am spirit awed by matter. Here is the magic of *object*, returned, once again, to front-stage center in human consciousness. Here is the amulet, the talisman, the wand, the totem, the icon, the idol. Here is the primitive shaman conjuring matter, the Egyptian preserving it, the alchemist purifying it, the priest transubstantiating it, the physicist transforming it, the drugtripper digging it, the schizophrenic trembling before it. The perennial wonder of matter. And how much difference is there between horror and wonder? Matter is always awesome, whether it delights or terrifies us. Do we have any choice in the matter? Perhaps choice is the key. Perhaps we may decide whether to grasp matter or tremble before it, whether to seize or yield to reality itself. But is there a way to renounce idolatry without being reduced to schizophrenic impotence?

This profound psychic state we call schizophrenia is a metaphor for life without metaphors. With its incoherence and absence of spacetime, it is a kind of plane of sheer existence, a country with no map or calendar, a world with no measure or substance. It is an utterly unfamiliar state of being. But is it unfamiliar, or, rather, unseen? Is it perhaps the bleak barren room we have camouflaged with the furnishings and comforts of home? Do we not add space, time, matter, and number to raw schizophrenic being to give ourselves form, support, motion, individuality; to provide a secure and habitable world?

Contemplating schizophrenia serves to attenuate the trappings of our reality and make them transparent. It loosens the stronghold that the cardinal metaphors have on our minds. It shows us the tentativeness of our world and its root in consciousness. It puts us back in touch with our creative potentials and active participation in reality.

The Leap into Faith

Becker's thesis is that the denial of death is the central driving force of life. Fated to exist as a paradoxical dual, a symbolic spirit imprisoned in a material body, a human strives, through countless heroic activities, to deny the raw facts of his or her existence and to create personal meaning and value. Whether one paints pictures, drives a truck, wages war, picks on one's kids, creates nations, builds bridges, or goes to a madhouse, it's all the same—impossible heroics. Becker calls these attempts at self-justification and validation *causa-sui* projects. Or, one may swear allegiance to a political leader, worship a guru, idolize a lover or a psychiatrist, thus projecting all meaning and value into another person in the classic pattern of psychological transference. Some of us even write books.

But is there no refuge far from this mad self-deceptive activity, no exit from the labyrinth of denial and avoidance? Are we simply doomed to an endless series of *causa-sui* projects and transferences? Must we go on forever kidding ourselves until brought abruptly to our senses on that terrifying, cold, gray dawn when we are dragged kicking and screaming to the gallows? Becker offers a hopeful answer to these questions, but it is complex and, as observed earlier, uneasy and ambivalent. He begins with Kierkegaard:

> One goes through it all to arrive at faith, the faith that one's very creatureliness has some meaning to a Creator; that despite one's true insignificance, weakness, death, one's existence has meaning in some ultimate sense because it exists within an eternal and infinite scheme of things brought about and maintained to some kind of design by some creative force. . . . But without the leap into faith the new helplessness of shedding one's character armor holds one in sheer terror.[4]

The giving up of all our escape and avoidance mechanisms leads to a state of terror. Reality is far worse, in this sense, than neurosis. Our protective armor, our screen against despair shields us from the horror of reality, from the truth about life and death. Therapy and psychological cure bring us up against the inescapable human condition from which there is no escape. What value does such truth have for us? Why should we pay such a price for it? Because it leads, in the words of William James, to "salvation through self-despair." Through despair, we destroy the vital lie and make room for a new life, for death and rebirth. From the bottom of the pit of despair, we may be resurrected by a leap into faith. Stripping off our defenses and dying in despair has no other purpose than to give us a new birth in faith, the ability to yield to a superordinate power and authority, and to abandon the need to create and justify ourselves through the *causa-sui* project.

But isn't this just another form of transference? How does surrendering to and placing all one's faith in God differ from idealizing and projecting all love and meaning onto a therapist? Does this not suggest yet another form of self-justification, in this case, of finding the meaning and guidance in one's life by acting out the will of God? Becker concedes, in fact, that this is true. He says that humans cannot do without some kind of meaning and purpose in life. The question for Becker is this: At what level of illusion does one seek immortality? What kind of heroics are the most sustaining, inspiring, creative? His answer: aim for the highest. Faith is transference to the greatest of all beyonds, to the most transcendent thing imaginable. We sacrifice the lower illusions to the highest of all:

> Out of the ruins of the broken cultural self there remains the mystery of the private, invisible, inner self which yearned for ultimate significance, for cosmic heroism. This invisible mystery at the heart of every creature now attains cosmic significance by affirming its connection with the invisible mystery at the heart of creation. This is the meaning of faith. At the same time it is the meaning of the merger of psychology and religion.[9]

Becker is quick to emphasize that this is no facile solution. We do not eliminate the terror, but rather we obtain through faith the "courage to renounce dread without any dread" and "to use anxiety as an eternal spring for growth." This is a remarkable idea—anxiety as the wellspring of creativity and trust—which Becker finds in Kierkegaard. We cannot escape the human condition. It is our bane. But by facing it without pretenses and illusions, the very terror it strikes in our hearts can become the source of new and ultimately sustaining metaphors and faith. Within our misery we find our courage, our strength, our cosmic heroics. This is a stoic and ironic approach to life. It is defiantly optimistic in the face of inevitable tragedy. There is no escaping the human condition, but there is the possibility of the creative and heroic use of it.

Personally, I find great encouragement in this. It fits in well with my own convictions that humans are the creators, or at least participant-creators, of their own world and that they must take full and honest responsibility for it. Becker insists on the necessity of these cosmic heroics and implies that people undertake them as moral and deliberate acts:

> if neurosis is sin, and not disease, then the only thing which can "cure" it is a world-view, some kind of affirmative collective ideology in which the person can perform the living drama of his acceptance as a creature. Only in this way can the neurotic come out of his isolation to become part of such a larger and higher wholeness as religion has always represented.[10]

And yet Becker states clearly that our cosmic heroics are not realizable goals, but only ideals, which ultimately depend on grace.

> Kierkegaard . . . placed himself between belief and faith, unable to

> make the jump. The jump doesn't depend on man after all—there's the rub: faith is a matter of grace.[11]

Grace, not will. Here is the core of the ambivalence for me. On the one hand, Becker says that the meaning of faith lies in affirming a "connection with the invisible mystery at the heart of creation." But on the other, one may do this affirming only if granted permission. Through faith we obtain the "courage to renounce dread," to "use anxiety as an eternal spring for growth." But only grace makes all this possible; without it we are left in despair. We strive for what can be attained only through divine intercession.

Becker, were he alive, could easily point out that there is no real ambivalence here: an ideal is never attainable; it is faith that gives us the will to keep striving. Seeming contradiction is the nature of the beast. We must strive for faith, but we do not receive it through striving. The ambivalence is perhaps mine and not Becker's: I refuse to yield, yet fear there is no other way. This is the dilemma of all who try to lift themselves by their own bootstraps.

Yet I continue to feel the ambivalence not only in myself, but in Becker's writing. His stoicism may well be interpreted as a patient fatalism about life without grace. But his vigor and courage belie such a diagnosis. He does not stand and wait; he charges forth. He is the champion of iconclasm, irony, and truth. He is Don Quixote, but with full awareness of his absurd plight. He is Camus. He admonishes us to maintain our skepticism while remaining "awe-filled creatures trying to live in harmony with the rest of creation" and "alive to the panic"[12] in it. He advocates a "creative myth" which does not simply allow a "relapse into comfortable illusion," but which "has to be as bold as possible in order to be truly generative," to give us the "courage to be,"[13] in Paul Tillich's honored phrase.

> No longer does one do as God wills, set over against some imaginary figure in heaven. Rather, in one's own person he tries to achieve what the creative powers of emergent Being have themselves so far achieved with lower forms of life: the overcoming of that which would negate life. The problem of meaninglessness is the form in which nonbeing poses itself in our time; then, says Tillich, the task of conscious

> beings at the height of their evolutionary destiny is to meet and vanquish this new emergent obstacle to sentient life.[14]

These are not the words of someone patiently awaiting grace, perhaps not even of one altogether convinced of its efficacy. But to guess the state of Becker's mind (or of his grace) is hardly any business of mine. It is far more appropriate to pay homage to him for his inspiring book, which is a valiant effort at vanquishing and absorbing into Being the obstacle of meaninglessness and nonbeing.

Nor is it possible for me to resolve the ambivalence over grace and will within myself or anyone else. It is too close to the source of action, motive, and purpose in all of us to be analyzed.

But I do find a great affinity between Becker's analysis of human illusion and my own. We are both interested in exposing illusions, but not in rejecting them altogether. Life without illusions is not possible, not even meaningful. What matters is to become more conscious of our myths and metaphors, to recognize that they are the only reality we have, and to learn how we participate in creating them. We must ask with Becker: what is creative projection? What is life-enhancing illusion? and seek with him and Kierkegaard the merger of psychology and religion, philosophy and science, poetry and truth in the yearning of the creature.

A Question of Life and Death

I am now prepared to take the idea of the cardinal metaphors as images of the current state of evolving consciousness and apply to them Ernest Becker's thesis of human activity and thought as an avoidance of death. In other words, to suggest that our projections of reality are largely formed in response to the human condition and the fear of death. But first I must pause to consider a question that arises naturally. How can the fear of death be a fundamental motivation behind our metaphors for space, time, matter, and number, when the ideas of life, death, and existence themselves depend on our spacetime concepts? Death is even conceived of, often, as the termination of consciousness in space and time as we know them. Furthermore, this chapter may seem to go against the prevailing grain of the book in arguing that psychological causes, grounded in the human physical condition, precede spiritual or cosmic forces. And this in turn is a contradiction, since the physical conditions in which we humans find ourselves must surely be a part, or a consequence, of the very cardinal metaphors they generate.

I have stated the whole matter in simplified logical terms to accentuate its contradictory character, but the real dilemma is a matter neither of logic nor even of precedence. As with the chicken and egg puzzle, there is no deciding which comes first. So many apparent contradictions in the mystery of mind and matter stem from the implicit assumption that mind and matter are separate to begin with. This whole book disputes that assumption, as the reader knows, and so the contradiction is somewhat artificial. Nonetheless, this chapter does shift emphasis from spiritual to psychological forces and the reader may wonder why.

The reason is that I feel both forces deeply within myself and sense each to be basic. This feeling is corroborated for me by many writers and thinkers, but the intuition is still my own. I believe consciousness and the view from within the subject to be inseparable from reality, and while I do not assume that each idiosyncratic view and whim is equally valid, I believe that a collective force of consciousness, in which we all participate, is synonymous with reality. At the same time, I know the fear of death to be so profound a force in the human mind that all attempts to diminish it by deriving it from something more basic leave me unconvinced.

Again, one may raise a logical objection: if consciousness creates reality, does it not create even death? Is not death itself an illusion, a metaphor? And again I must protest that it is not a matter of precedence. Consciousness neither precedes nor follows reality; they are one and the same. Yet here I sense a partial resolution of the dilemma. It may be that some metaphors are more fundamental than others. Indeed, I have said as much of the cardinal metaphors. But which are the most fundamental? In some hierarchy of metaphors, could any be more essential than space, time, matter, and number? If pressed, I might say that existence/nonexistence takes precedence even over the cardinal metaphors, that being itself must come before any of its characterizations.

Recall the opening line of the Hindu *Creation Hymn* quoted earlier which invokes some primal state:

> There was not non-existent nor existent.

However difficult to comprehend, there occurs in primal cosmologies, such as the Hindu, the idea that prior to any manifestation is a cosmic state of pure unalloyed being—timeless, spaceless, featureless, prior somehow even to existence. Such an unimaginable state expresses its first manifestation, its first breath or rhythm, so to speak, as an existence in, against, or out of something. The metaphors associated with this first primal existence and its complement, nonexistence, might indeed be expressed even within cosmic consciousness as life and death. These in turn might be the stuff out of which the cardinal metaphors are fashioned. Out of primal being comes existence/nonexistence or life/death, and from this emerges space, time, matter, and number. In this sense, life and death may be seen as metaphors of consciousness and yet as forerunners of derivative metaphors.

But I cannot back up these speculations with the same strong convictions I hold on the primacy of consciousness and death. I offer them as a metaphysical rationale to answer the objection this chapter raises. Ultimately, I believe there is no explanation for such matters. The one I've given implicitly assumes an evolution in time behind the idea of precedence and, therefore, that time is prior to, or at least simultaneous with, being. It suffers from the fallacy of analysis—the attempt to divide the indivisible into parts. It seems impossible to begin from nothing and make something. At least, it seems impossible to comprehend or describe this. Yet it may be the simplest thing in creation.

Secret Agents

Ernest Becker shows that the flight from death is the central motivation of human activity and creation. By means of *causa-sui* projects and transference we give value and meaning to our lives and try to overcome our absurd animal fate. Now, the creation and signification of meaning and value is the fundamental role of metaphor. The cardinal metaphors reflect our deepest human yearnings and needs, and deny the meaninglessness of chaos and death. Space is a metaphor for being, for existence, for standing out against some background or medium. Astrological space provides a feeling of relatedness to the cosmos. Time symbolizes persistence and endurance, with hints of immortality. Its linear regularity protects us from the unexpected and the cataclysmic, while tempo and rhythm reflect the pulse of nature and the elemental acts of creation and destruction. Matter is our metaphor for stability and security. Substance gives us something to hold on to, and inertia protects us from arbitrary flight and downfall. And number represents unity and diversity. It hints at our common root while keeping us and all things separate and denumerable. Let's look at some of this in more detail.

In thinking about the ground of our being, it is very difficult to separate space, time, matter, and number from one another. We saw earlier how entangled and interdependent the cardinal metaphors are. As we try to relate them to the very notions of our identity, existence, duration, etc., they blend in our minds all the more. I begin with space, recognizing all the while that the choice is arbitrary and that dissection is really impossible without killing the organism.

The states of consciousness which we may call mythic, astrological, perhaps schizophrenic, are characterized by an organic, interrelated, fluid, even chaotic experience of space. By contrast our modern spatial metaphor is the epitome of distance, extension, and articulation. Modern space is rigid, divisible, compartmentalized, empty, lifeless, incompressible, impenetrable, external, remote, and of infinite reach. It is the background, par excellence, from which an identity may emerge. It is the medium in which we are singled out and localized as unique and distinct entities. Our very sense of existing as separate selves is intimately tied up with our spatial metaphor. To experience oneself as an individual defined in contrast with all other things requires some such construct as space. To exist at all is to *stand out*, and to stand out means to have something to stand out *from*. Space is that from which we stand out, in which we exist, without which we cannot be. Our sense of self is grounded in an awareness of our existence, and that existence is experienced in space.

To the extent that our modern spatial metaphor intensifies our sense of distinctness, uniqueness, and unconnectedness, to that extent the *existential* problems of life, those that concern our *standing out*, have become more pronounced and dramatic. To transcend space and time cannot properly be conceived of as an alternative form of existence, but rather as no existence at all, and this is how we view death. The testimony of mystic seers, saints, and psychedelic travelers supports this view. They tell of an omniscient or oceanic feeling but without a sense of individual localized identity. Such descriptions appear to involve consciousness, although at its highest levels (variously called bliss, samadhi, nirvana, heaven) we are told that there is no longer anything to be conscious of. Subject and object are no longer distinguishable. Words fail. To speak literally of consciousness and even of existence in these contexts is folly. People do return from such states and tell us about them. But how are we to treat their words? Only as metaphors, surely. In such states even existence is a metaphor.

To say that the self exists in space is more or less a way of talking about and conceptually relating three things—self, existence, and space—which are really inseparable. It is a metaphor of our experience, of our life. Death is a state of nonexistence which means no more standing out of a self, no more being in space. Construing ourselves as existing in space is a kind of insurance against nonexistence, against death. Whatever death may be, we sense it to be a denial of existence as we know it. There may be some inchoate merging into the chaotic All, but the self as such will no longer *be* in extended space. Space is one of the guarantors of our existential experience. We create it, project it (or participate in these acts) to ward off death.

The other profound metaphor that acts as a guarantor of our *continued* existence in space is, of course, time. Our existence as distinct selves is threatened not only by possible dissolution into space (or collapse of space) but by abrupt termination in time—the horrendous cutting off of life so perfectly symbolized by the guillotine. In fact, the finitude of life seems the greater threat, so that most of our wishful thinking concerns immortality rather than, what shall I call it, ubiety. Perhaps space is so deeply ingrained in human consciousness that even the fear of its cessation cannot be properly formulated by the mind as a distinct emotion. Yet certain fears and nightmares, such as being buried alive, losing sensation and motility, and being devoured or dissolved, do suggest, however inarticulately, the loss of spatial existence.

As the cardinal metaphors merge imperceptibly into one another, who can say whether our sense of separateness is more spatial than numerical. Or our body awareness more material than spatial. Or our experience of motion and rest more inertial than temporal and spatial.

But whatever the reasons, our fears of death seem to focus most on the human sense of time. Our dreams of immortality are rarely preoccupied with what form in space we shall take so long as we persist. We eschew aging and decay, seeking always youth and immortality. It is not clear which is the disease and which the symptom. Do we celebrate youth and scorn old age on aesthetic grounds or because the old remind us of our own deaths? We visualize our persistence in terms of an ordered continuing sequence of events in time. A blending of time, continuity, and causality is buried deep in our immortal longings. We want to endure forever, we want no abrupt changes, and we want to be able to predict the future—the perfect ingredients for avoiding death. Duration in time separates us from death, if only temporarily in the all-important now. Continuity and causality give us security and a sense of permanence.

The very words that imply lasting existence reveal our rationalizations. *Duration*, *endure*, and *durable* contain the stem *dur*, derived from two different, but similar Latin words: the verb *durare*, to last or endure, and the adjective *durus*, hard. Their similarity, even in the Latin, is significant. Duration in time implies persistence as well as extension. The latter meaning is usually associated with the spatialization of time, so common in quantitative applications. But even in everyday language we talk of a spread, a lapse, a period of time, and of tightening our belts for the duration, in World War II parlance. We are all familiar with time as interval. And, of course, persistence is just beneath the surface of duration. What is it that persists or lasts or endures?

Civilizations have been fascinated with
lasting concrete symbols of immortality.

Why, whatever is hard, *durus*. Hard objects endure the ravages of time and give us a feeling of permanence and immortality. Why all this preoccupation with hardness, toughness, sturdiness; with clenched fists, rigid muscles, and erections, if not because deep down within us they symbolize immortality?

Our preoccupation with endurance and hardness suggests some deep unconscious human longing for eternal life. Through the ages, civilizations have been fascinated with lasting concrete symbols of immortality: the rock, the monolith, the phallus, the temple, the tower, the skyscraper, the jet, the rocket, all made of the hardest, most enduring substances. But we have also dramatized our immortality in the conflict between duration and persistence on the one hand and variation and change on the other. The stage was set for the great struggle between the two views nearly three millenia ago by Heracleitus and Parmenides. Is change or permanence more basic, they asked. Plato and Aristotle contributed to the debate and still affect our thought. It is the permanence of Plato's ideal forms beckoning to us through Aristotle's substance that we feel today in our modern physical laws and conservation principles. These laws state that the quantity of certain forms or substances neither increases nor decreases but persists in time. The form may change but the amount endures. Electrical energy, for example, may be converted into heat, but the total quantity of energy remains constant; none is lost or gained. We can see our deepest subjective yearnings abstracted, objectified, and canonized in the physical sciences. The conservation principles of physics are its most widely believed and sacrosant laws: conservation of energy and mass, of electric charge, of momentum and spin, to name some of the more familiar and inviolable examples. Through them, the theme of immortality and endurance is singled out as the most important characteristic of the structure of physics. Of all the myriad possible forms scientific theory might have taken, the one we unconsciously chose emphasizes and enshrines persistence in time—immortality. If our buildings, artifacts, and masterworks are not of sufficient endurance, we shall imprint our mental structures in the very fabric of the cosmos. We marvel at the intellectual order of the universe. But how do we recognize it? Whose identifying stamp do we find out there?

When we peer through our telescopes, is it not our own immortal yearnings we see?

We have enshrined our immortality in a cathedral of space and time, but it has become our prison fortress. Its unyielding walls form the space that confines us, and its ebbing time is our life sentence.

Chapter 8.

Participation without Representation

Our imprisonment in spacetime is not yet at an end, but we may be eligible for parole earlier than we imagine. And while we're waiting, we might just as well make our cell a little more comfortable for ourselves. Gradually, as we come to understand our creative participation in the physical world, we shall realize that where and how we live is largely up to us and that the fate of the world—its ultimate perfection or downfall—is our responsibility.

If we adopt a dualistic framework (as I do not), human morality takes on a kind of indirect, artificial character. We feel obligated to someone or something else. Our moral involvement becomes legalistic rather than personal.

We assume an external world exists out there, which is independent of our human minds. In this dualistic world we are, at best, lodgers; at worst, poachers. It isn't our stuff. We don't have the direct responsibility of the owner and manufacturer. We beg, borrow, or steal the goods from which we make clothing, symphonies, and atom bombs. Our primary responsibility to each other (and to a few of the "higher" creatures around) is to use our hand-me-down playthings to help, rather than harm, each other. Beyond that we have no obligation. We discard and despoil our borrowed toys.

During my childhood in an uptown Manhattan neighborhood, I recall hearing a secondhand clothes dealer who would periodically push his old-fashioned cart through the streets and cry out, in a singsong refrain, "I cash old clothes. I cash old clothes." To me, this man was one of those unfathomable mysteries associated with the complex adult world. I was never quite sure what he was saying or what he did with old clothes. But his erratic appearance every now and then would first jar and then sooth me with an uncomfortable familiarity. While I would play safely indoors, the funny, disquieting man was out there in the street ministering to some grownup needs, returning value for castoff goods.

This apparition from the past is somehow the perfect metaphor for the dissociated responsibility we feel today toward the physical world. If we can turn a pretty penny on our tattered, old-fashioned garments, we gladly relinquish them. We didn't make them to begin with, and now that they've outlived their usefulness, we're only too glad to let the I-cash-clothes man take the responsibility off our backs.

But if we reject dualism, human responsibility becomes genuine, deep, and pervasive. There are not two separate worlds, there is one. We participate in it, and we help to make and sustain it. We are not renters or squatters, but the owners. We make not only the clothing but the very fabric of matter from which it is woven; we create both the sound and the time that make music possible; we manufacture not only the bombs but the raw atoms from which they're made. We are responsible not only for other creatures but for every *thing* in the cosmos, for the very spacetime in which all exists.

This may sound preposterous, but I believe it is the most reasonable consequence of the inevitably subjective basis of space, time, matter, and number. The cardinal metaphors are human creations of value and meaning. We are the authors of the iconography of the physical world. Length is not something that exists in the outside world independently of us. It is part of space itself. It is a subjective value judgment like beauty or justice. We ourselves have made it. If we do not accept this, we must explain how length can exist in a world devoid of human consciousness. We cannot escape the responsibility for what is ultimately subjective—the physical world. No more cash for clothes.

Ecological Consciousness

There are abroad in the world today several signs of our growing sense of responsibility, of our evolving consciousness. We hear talk of *ecological consciousness*, for example. This notion carries with it an assumed responsibility for the very air we breathe and the water we drink. We are aware, as we never have been before, that we are fouling and despoiling our own environment. Simple animals know better than this, and at last we are beginning to learn. But more is involved than housecleaning. We claim we are preserving the environment to save ourselves. This is surely true. Yet we talk of a responsibility to the planet itself, even in the event that humans do not survive. We are awakening our sense of connectedness to nature and the physical world. (I should call it organic and alchemical consciousness.)

In this we can learn much from the American Indian. Most of these people experienced a deep sense of connection to and interrelation with their surroundings. They could no more conceive of polluting a river than we can of poisoning our bloodstream. For them, the distinction between an inner and outer world is not especially meaningful. Rocks, trees, and streams are not seen by Indians as other and external, but as part of a spiritual continuum to which they, themselves, belong. Consciousness and life pervade not only human beings, animals, and plants, but rocks and minerals, water and air, sky and stars as well. This is no simple anthropomorphism. It is, in fact, presumptuous to judge such an elevated state of awareness and being in terms of our contrived, modern theories of society and behavior. This is rather a different consciousness. Nor can the Indian's feeling of shared life and mystery be reduced to any simple common denominator. Black Elk,[1] despite his exquisite sensitivity to the magic and spirit of the natural world, failed to comprehend what white people find so hypnotic about gold. More cash value.

It is precisely the failure to feel our connection to the physical world coupled with our reduction of all quality and meaning in nature to some simplistic scale of values that has resulted in our deep environmental problems. The use of fossil and nuclear fuels and other dangerous and polluting sources of energy are invariably justified on the basis of productivity, efficiency, or profit. Can the function of the kidneys, the lungs, or the brain ever be judged in comparable terms? The value of such things cannot be reduced to dollars or percentages. Neither can the use of energy and matter.

If we allow our consciousness to sink into and contemplate the materials and fuels we use so glibly today, we shall understand from within the harm and filth inherent in so many of them. Just as our senses can warn us of the inherent danger in fire, so our consciousness can make us become aware of the invisible, insentient threat within radioactive matter. Fire may be harnessed, but only at a price—the price of smoke and ash, poison gasses and excessive heat. We claim to have harnessed nuclear energy. But at what price? The smoke and ash of fire are harmful enough, but if handled properly and not produced too rapidly, they may be recycled and reabsorbed into the environment. But nuclear ash and smoke remain dangerous and destructive for tens, hundreds, even thousands of years in some cases. And to the extent that radioactive wastes are recycled and absorbed into the environment, they become more of a menace, not less. In good conscience, we are obligated to isolate and store these wastes for centuries upon centuries. These substances, in the quantities produced today, are incompatible with life, consciousness, and organic order in the universe. As human consciousness reclaims its former powers and sensitivities, and enlarges upon them, we shall know these things in our bones. We shall know better than to despoil our own nest and to bequeath such a deadly legacy to our children and their descendants.

The modern advocacy of solar, wind, and geothermal power, of organic agriculture and small family farms, of environmental protection and the preservation of wilderness, of natural foods and vegetarianism, of the rescue of endangered species, of campaigns for clean air and water, and of intermediate and appropriate technologies foreshadows, I believe, much more than a new political and economic movement. It heralds a genuine change of consciousness, a metanoia, a regeneration of the human connection to the environment, a birth of new metaphors. It is no wonder that advocates of these new movements often find themselves wrongly understood and interpreted, even by well-intentioned adversaries. The opposing groups are people with different iconographies, languages, and perceptions: shamans arguing with engineers or surrealist painters with abstract expressionists.

Reading the environmentalists from Thoreau to the present, one feels more than rational argument or the mere presentation of a case. One feels a sense of alarm, an awareness of a spreading corruption within the body. There is a pathos, too, a sadness for the loss of each dear friend and loved one. One reads not the abstract tabulations of morticians and actuaries, but the personal anguish and grief of a mother for her dying children, her own flesh and blood. These are cries from the heart, from a mind that knows and understands its own organs and the delicate balances that sustain them. This is the voice of a new consciousness.

Holistic Medicine

Among the greater successes of post-Enlightenment consciousness and science is modern medicine. Its achievements in the art of surgery and in conquering infectious diseases, the ancient plagues and scourges of humanity, are by now legendary. Hardly a man or woman alive today (excluding the third world, unfortunately) does not owe a debt to modern medical therapies. And yet the great fortress of modern medicine is beginning to show signs of weakness. The direct treatment of symptoms through drugs and surgical intervention has met some powerful foes in the cardiovascular and pulmonary diseases and cancer. Much modern medical practice is characterized today as too clinical and inhuman and sometimes, as in the case of the dying, as more harmful than beneficial. The financial and organizational problems involved in dispensing medical services to the population at large are beginning to get out of hand. Medical therapy is itself creating dangers to health as in the case of iatrogenic diseases. Finally, medicine is frustrated by the fact that nutrition, food additives, smoking, sanitation, pollution, and the stress and strain of daily life all create medical problems that are far better handled through prevention than cure.

In response to these failings of traditional modern Western medicine, and as a further outgrowth of the new consciousness, a number of new approaches (and, in some cases, rejuvenated and modified old ones) to health and illness have attained importance in the past decade. These holistic therapies include, among others, homeopathy, acupuncture and Chinese and Eastern medicine generally, spiritual and psychic healing, nutritional and dietary therapies, and various psychological/psychic therapies with a strong bearing on somatic medicine, such as bioenergetics, gestalt therapy, and visualization or mind-over-matter techniques. These are not crackpot therapies (always a danger in times of rapid evolution). They have many successes, advocates, and serious practitioners. They are capable of improving and enlarging Western medicine by bringing to it their holistic, organic point of view.

Homeopathy, for example, attempts to treat the whole person, rather than specific symptoms. The homeopathist contemplates a patient's entire history and constellation of symptoms and maladies in an effort to arrive at an intuitive synthesis, a unique characterization of the patient's state of health. On that basis, a drug is prescribed. The medicine proceeds to penetrate the various levels and stages of the patient's past illnesses, which have culminated in the current symptoms. In the course of treatment, earlier problems, long suppressed by Western medical practice or naturally dormant, may recur as the path of the illness is traversed in reverse, until finally the drug reaches the source of the problem and corrects the basic imbalance. Homeopathic philosophy, which significantly stems from medieval alchemy, is based on the ancient notion of order and harmony. It attempts to treat illness subtly by divining and then correcting disharmonies deep within the body and the psyche. It does this through the use of substances that resemble or have an affinity for the imbalance. This contrasts sharply with the Western approach in which the more outward symptoms are treated first and with the strongest counteractive agents.

Homeopathy assumes as a matter of course that mind and body are inextricably related, that consciousness plays an active role in treatment, and most significantly (and this homeopathy shares with Eastern medicine, generally) that a person's medical history is all of a piece, like a convoluted, coiled snake, that it exists in some sychronistic realm and must be conceived of and treated whole.

A second example of a new consciousness therapy is the visualization/meditation therapy used in the treatment of cancer by Dr. O. C. Simonton of Fort Worth, Texas. Dr. Simonton was trained and certified in traditional Western medicine, but his search for a more effective treatment for cancer has led him to combine Western therapies with a mind-over-matter approach adapted from Eastern meditative techniques. As a supplement to the standard Western surgical, radiative, and chemical treatment of cancer, Dr. Simonton's patients undergo a training program in which they try to take some control of what is happening in their bodies. They record and listen frequently to tapes that encourage and reinforce a positive and constructive attitude. This helps the patient to overcome the usual feelings of fear and frustration associated with being a helpless cancer victim. Central to the approach is a meditative technique in which patients fantasize seeing themselves fight the disease at the level of tumors and cancer cells. In their imagination, they may picture their own white blood corpuscles attacking and engulfing the cancerous cells. Or perhaps, tiny animals are seen eating the malignancy, or some substance in the blood is seen to dissolve it. Any fantasy concoction that is easy to visualize can do the trick.

Through these methods, Dr. Simonton finds that the patient's immunological system is strengthened in fighting cancer, that a larger than normal percentage of patients survive and recover, that the average life-span of cancer patients increases, and, perhaps most important, that the attitude and outlook on life, even of dying patients, is greatly improved. Western medicine can benefit enormously by incorporating these holistic therapies. Whether it can survive their organic, nonmechanistic philosophies of mind and body is another matter.

A Holographic Universe

Finally, we shall note a modern attempt to take serious account of consciousness in a suprising marriage between physics and psychology. It is the new holographic theory of mind and matter professed by Karl Pribram and David Bohm.[2] Pribram, a psychologist, and Bohm, a physicist, point to an essential feature shared by both the inner and outer worlds—that every part of the whole contains or implies the whole. Mental functions are not localized in the mind, for example, but are diffuse: memory is everywhere, not spread out in the brain. Similarly, information about the whole cosmos is contained in each local region: every part of the universe knows, so to speak, what the rest of the universe is up to. The Pribram-Bohm model for this part-knows-all character of mind/matter is the hologram, a special kind of image-producing photographic transparency. An illuminated hologram produces an amazingly realistic image of an object in three dimensions. If one moves one's head while viewing a holographic image of a car, for instance, one can see a top, bottom, and side view, just as if the car actually existed in space before one's eyes.

What is so important to Pribram and Bohm is that if only a portion of a hologram is illuminated, one still obtains a full image of the photographic object and not just a part of the image, as would be the case with a photographic slide. The hologram, unlike a photograph, looks nothing like the object it represents. Instead of a two-dimensional projection of the object, it is a pattern representing the state of the light waves coming from the object. When this pattern is illuminated, the original light waves are reconstructed in reverse and projected back to form an image of the object in space with all its three-dimensional properties. And since any point on the hologram receives light from all points on the object, each part incorporates information about the whole. The image formed by a portion of a hologram is somewhat less detailed than that from the full hologram, but the point of the Pribram-Bohm analogy is that a hologram contains information about the whole in each of its parts.

In the new theory, we can say that mind and matter each display this property because mind and matter are one and the same. This part-equals-all property is the basic nature of the aspatial, atemporal realm where all is one.

Pribram and Bohm continue to work on the details of this exciting theory. It will demonstrate, I believe, that consciousness and physical reality are different representations of the same thing, and that our conventional concept of space and time, which fails to take into account the part-equals-all character of the hologram analogy, is at best a useful though quite incomplete metaphor for our conscious universe.

The Cardinal Virtue

But what direct evidence is there, you may ask, of your own creative act? You do not feel yourself creating and sustaining the objects around you. If human participation in the physical world is truly becoming more conscious, how can one become aware of it? The difficulty here is not the foreignness and infrequency of our creative participation, but its intimacy and pervasiveness. It is so much a part of our perception, thinking, and conceptualization that we are almost totally unaware of it. To understand this, we must take a brief look at the psychology of perception.

Owen Barfield, in discussing collective representations, our commonly shared metaphors for the physical world, demonstrates how, through perception, we participate in constructing these metaphors:

> Perception takes place by means of sense-organs, though the ingredient in it of sensation, experienced as such, varies greatly as between the different senses. In touch I suppose we come nearest to sensation without perception; in sight to perception without sensation. But the two most important things to remember about perception are these: *first*, that we must not confuse the percept with its cause. I do not hear undulating molecules of air; the name of what I hear is *sound*. I do not touch a moving system of waves or of atoms and electrons with relatively vast empty spaces between them; the name of what I touch is *matter*. *Second*, I do not perceive any *thing* with my sense-organs alone, but with a great part of my whole human being. Thus, I may say, loosely, that "I hear a thrush singing." But in strict truth all that I ever merely "hear"—all that I ever hear simply by virtue of having ears—is *sound*. When I "hear a thrush singing," I am hearing, not with my ears alone, but with all sorts of other things like mental habits, memory, imagination, feeling and (to the extent at least that the act of attention involves it) will.[3]

In other words, what I call matter is neither what causes my sensations (presumably atoms and electric fields do that), nor equivalent to my sensations (which are a complex of tactile impressions and visual images). Matter is something I construct mentally out of my sensations. This conversion of pure sensation into a perceived object, Barfield calls *figuration*.

In *The Tacit Dimension*,[4] Polanyi employs a beautiful illustration in which we can actually *feel* the figuration process developing within us, if we place ourselves in the shoes of a blind person who is learning to "see" through a cane. At first, you feel only the impact of the handle of the cane against your palm and fingers. It is a new and unfamiliar sensation. Gradually, you become accustomed to using the cane as a probe to feel your way around. The immediate sense of the impact on your hand is transferred to an awareness of the tip of the cane as it taps on the objects you are exploring. With further experience, you lose even the awareness of the tip of the cane and begin to sense, and then see in your mind's eye, the object probed by the cane. Eventually, you build up a mental picture of your environment and the space around you. Through the interpretive act of figuration, meaningless sensations are transformed into a meaningful construction—a visual space, in this case.

Here is the very process of constructing a metaphor through imagination. If I am blind, I evoke a space all about myself by virtue of the sensations in my hand. I displace and project these sensations into a surrounding distant environment and eventually become completely unconscious of the tactile impressions I build it out of, just as I am totally unaware of the light striking my eyes when I see things around me in space. Combining these illustrations, we can imagine, too, how one learns to hear a thrush singing. I, as a novice bird watcher, for example, will not at first recognize the call of a thrush. I must attend carefully to the auditory signals I get from different birds. Through careful practice, I will slowly learn to equate a bird with its song and to distinguish different birds from one another. Eventually, as an experienced bird watcher, I will hear and immediately recognize a thrush with no conscious effort. I will not dwell on the sounds I hear, but will transform them instantly into the image of the bird, just as we do not dwell on the letters and words as we scan a page, but rather read the meaning.

There is an active imagination in the mind, of which figuration is a part, by means of which we create the world of our experiences. We have for so long buried it under our rational, intellective faculties that we have fooled ourselves into thinking it does not exist, but it continues to flourish, with or without our notice. Goethe was acutely aware of it and sought to make it the basis of science as well as art. Coleridge placed it at the core of his whole philosophy. Rudolph Steiner put it into practice in a radical new approach to medicine, agriculture, education, and spiritual enlightenment. Einstein felt it in the very creation of a new science. Nor is it the exclusive possession of the genius, nor the private property of the individual. Our creative, intuitive imagination is something we all share and through which we collectively project our reality.

For me it is not important whether we view ourselves as participating *with* God or as the sole authors of the Creation. It is certainly not our private individual power, but one we share with the community of the cosmos. In any case, such a question can only be posed (although never answered) from within an assumed dualistic framework in which God and human, like mind and body, are conceived as separate. Nor is this claim of creation idolatrous, for it is not the creation of images but the worshiping of them that degrades us. Becker shows us that we must make metaphors, but we need not do so in blindness and idolatry; we need not be their captives. The essential issue is for us to create consciously and morally, to participate *with* representation.

It is for us to decide whether it shall be our dreams or nightmares that will come to pass. No horror imagined in the human mind has been too great to prevent us from realizing it in the course of history. Are we not capable today of releasing the holocaust of final doom? The hidden potentiality of the human mind and our collective imaginings is staggering. It is for us to decide to use that potentiality, that creative imagination, for the betterment of the world, not for its destruction. The interlocking, indeed the unity, of aesthetics and ethics is inescapable.

In Owen Barfield's paraphrase of William Blake, "Imagination is the cardinal virtue, because the literalness which supports

Shall it be our dreams or nightmares that will come to pass?

idolatry is the besetting sin of the age which is upon us."[5] Our little cardinal returns once again. And whether thrush or cardinal, we'd best understand who it is that brings him around to sing and cheer us up and who is reponsible for his well-being (*being* indeed!).

Chapter 9.

New Metaphors for Old

It's very much more exciting to discover we're on a ball, half of it sticking upside down. It's spinning around in space; there's a mysterious force which holds us on it; it's going around a great big glob of gas that's burning by a pure fire that's completely different than any fire we can make (though now we can make that fire—nuclear fire).

But that's a much more exciting story to many people than the tales which other people used to make up who worried about the universe—that we were living on the back of a turtle or something like that. They were wonderful stories, but the truth is so much more remarkable. So what's the pleasure in physics to me is that, as it's revealed, the truth is so remarkable, so amazing. And I have this "disease." And many other people who have studied far enough to begin to understand a little of how things work are fascinated by it. And this fascination drives them on to such an extent that they've been able to convince governments and so on to keep supporting them in this investigation that the race is making into its own environment.

If any one remark could serve as a target for the polemics of this book, the above statement would be my choice. It was made over public television in the United States a few years ago by a world-famous Nobel laureate in physics. While I refrain from giving it, his is a household name for those even remotely conversant with twentieth-century physics. If I single him out, it is not as an individual, but as the spokesman of an attitude typical of many physical scientists who believe they are engaged in finding the "remarkable . . . amazing . . . truth . . . as it's revealed."

As an individual, he could hardly be more exemplary. He is greatly admired for his many original contributions to physics. His devotion to science education and popularization bespeaks a strong moral commitment to humanizing science. Furthermore, his is no literal mind objecting to allegory or symbol. In his statement, he appreciates and pays tribute to myth. His rejection of ancient tales in favor of modern truth is based, in no small part, on matters of taste and aesthetics—"They were wonderful stories, but the truth is so much more remarkable." He seems strongly motivated by a spontaneous childlike sense of awe and wonder about nature. He easily and naturally communicates his sense of enthusiasm and joy, and has succeeded in inspiring countless students and colleagues over the years through his books, articles, films, and lectures.

In recounting the character and accomplishments of this well-loved and respected scientist, I make it very difficult for myself to criticize him. And yet I must, for I believe that the philosophy he expresses is idolatrous and harmful. The fact that the remarks were made in an informal conversation makes them even more important. For I am concerned with the way many scientists characteristically view the value and significance of their own work, which is best revealed in the unguarded comment. They see themselves in pursuit of an elusive but approachable truth. In many scientific publications and popular lectures, we find references to pushing back the frontiers of knowledge, and the heroic mission of scientific discovery, or our obligation to the truth. When pressed on these matters, most scientists will concede that all physical theories are tentative and approximate and that there is no clear conception of physical reality in modern quantum theory. But such disclaimers are made in the interest of fair play, so to speak. They do not characterize the way scientists behave and think in their everyday activities of research and teaching nor in their conversations with each other.

The fundamental idolatry in these views is their implicit assumption that an external physical world exists as an objective reality independent of the human mind and that the business of science is the discovery and description, not the creation, of that world. Human inventiveness, imagination, and taste may be involved in the formulation of scientific theory, but as a matter of form, not of content. Classical mechanics, for instance, can be framed in alternative ways. One may use Newton's original formulation of the laws of motion or Hamilton's principle of least action. There is leeway in the expression of the laws of nature which permits an element of human creativeness to enter in. But we must not let this fool us into thinking that we are responsible for these laws. The acid test of any scientific theory is, first and foremost, its agreement with the *facts* of the physical world. It is empiricism, not aesthetics, that is the backbone of science. Any theory, no matter how beautiful, will be rejected as soon as it is found incapable of corroborating the facts of nature.

Admittedly, this view is extreme and somewhat naive—that of a straitlaced logical empiricist. It is rarely held today in such a pure form by any philosopher or contemplative practitioner of science. But in my view, many of the modifications and reformations of the basic empiricist view are in the nature of hedging one's bets. They are changes in degree of restrictiveness; they are quantitative, not qualitative. They admit to the mathematical nature of many things in science and that many concepts and quantities in physics do not correspond with anything directly verifiable in nature. But the hard core of these views is that somewhere, somehow, sometime, any valid theory must be made to produce *realistic* numbers that correspond with the directly observable and measurable quantities in the physical world.

Now it is not my intention to present a full-blown argument against this point of view. Many writers and philosophers have already done that. One can read Poincaré[1] to see how facts are selected to fit in with one's intuitive sense of the harmonious order of nature. Polanyi[2] demonstrates the essential subjective and *tacit* component in all scientific understanding and theorizing. In Kuhn's[3] work, we see how the currently held scientific *paradigm* (belief system) strongly influences the interpretation of experimental data and the acceptance or rejection of new theories. Holton[4] documents the pervasiveness in scientific thought of *thematic* elements—preconceptions, such as religious and other dogmatic beliefs about the nature of the universe. Brown[5] presents a full-scale critique of logical empiricism and exposition of the new philosophy of science. And there are many others, such as Hanson, Barfield, Toulmin, Koestler, Roszak, Pauwels and Bergier, von Weizsäcker, Mumford, Bateson, and Eiseley, who deal with important aspects of the subjective and representational in the physical sciences.

My principal objection to the statement that opens this chapter is to its ironic unwillingness to recognize in our present scientific world view the same storylike quality as there is in the tales of those who thought they lived on the back of a turtle. This irony is very sad. For in responding to the excitement, wonder, mystery, and amazement of the cosmos, but in falsely attributing them to some objective truth, we betray the human spirit they reflect. We fail to sense and experience our own act of participation in nature.

One of the greatest creative achievements of the human mind, modern science, refuses to recognize the depths of its own creativity, and has now reached the point in its development where that very refusal blocks its further growth. Modern physics screams at us that there is no ultimate material reality and that whatever it is we are describing, the human mind cannot be parted from it. And yet we turn deaf ears to this profound cry. True, we don't know how to accommodate this idea: we don't know how to modify or enlarge physics so as to assimilate the mind which it has for so long exiled from its territory. But that is a poor excuse. We are acting out of fear and ignoring the moral responsibility inherent in our creative act. It is time to acknowledge and exercise our own redemptive powers. And there *are* clues all around us—in psychology, in philosophy, in the arts, and elsewhere. The first responsibility is to help others to see, so that we may all grow together. Education, this time without indoctrination.

Teaching Physics as Metaphor

Since I believe education is the starting point in exorcising idolatry, I shall make a few suggestions for teaching and learning, both formal and informal. Later in this chapter, I'll offer some exercises and puzzles that might be useful to students and practitioners of science as well as laypersons. But first I want to talk about ways of teaching physics in the more formal setting of the classroom. I'll stick to physics since that's what I know best and have taught the most. But I believe my remarks can also be extended to other scientific fields. In any case, I will not be treating physics in isolation from her sister sciences.

I shall also ignore questions of new methodology in teaching. Throughout my teaching career, I have advocated and practiced methods that give greater responsibility and freedom to students. I have allowed them to select subject matter, grade themselves, work independently, plan their own programs, apportion their own time, and so on. I believe such practices have a place at all levels of education and are especially valuable in helping students avoid indoctrination. But here I am primarily concerned with the content of metaphorical physics.

THE HUMANITIES APPROACH

It is well known in high school and college education that teaching required physics courses to nonscience students can be very frustrating. Most people, including students, concede that science is an essential part of a liberal education, but that doesn't make it any more palatable. Physics often proves distasteful to students, even when a special effort is made to provide good instruction and an interesting course format. I believe the reason is that few students ever see the human aspects of physics. They may, if they are lucky, have an instructor who is genuinely interested in his or her work and in teaching it, and who can communicate a sense of his or her own excitement and commitment. But in spite of being impressed by such instructors, few students can figure out why anybody would ever get excited about such dry stuff. They simply cannot sense, as an inward experience, the meaning and value that physics has for that small class or incomprehensible people called physicists.

Throughout this book, I have tried to illustrate that questions of meaning, value, beauty, life, and death are an intimate part of the concepts and ideas of physics and of our picture of the physical world itself. What motivates physicists, consciously or not, is the same thing that motivates artists, philosophers, theologians, politicians, psychologists, and in a less obvious but equally urgent way, busdrivers and chefs, viz., the quest for meaning and value. The trouble is that this is never quite visible in the classroom, because most physicists long ago rejected such concerns from their thoughts about science.

It is essential that teachers of physics examine their own motivations and fears, secret beliefs and wishes, fantasies and dreams. Somehow they must relate them to their drives and interests in physics and life. And they must trace them in the development and history of physics, and present them in the classroom. A personal confession may not be called for, but some kind of public distillation of the heady elixir of emotions, fears, and motivations is needed if students are ever going to understand the human concerns that physics deals with, however subliminally.

In a literature or humanities course, students can easily sense the human motives, fears, desires, and other feelings that lie beneath the surface, and can identify with them. But in the cold objective realm of physics for which they have no natural affinity and where they find no warmth, what flesh or emotion is there to identify with? The humanities are concerned with questions of existence, meaning, value, and beauty—matters which all of us feel to be essential, integral, human. It is ironic and tragic that we have come to imagine that physics is not deeply motivated by these same human issues. Ironic, because the inchoate yearnings and questions that moved early people to peer into the skies for understanding and meaning have now evolved into a discipline that seems to have rejected its own origins and inner life. Tragic, because we doom ourselves to alienated existence by refusing to recognize our own needs and our power to create the meaning that sustains us.

In the modern scientific period, we have made a cult and a fetish out of this rejection and refusal. The father of modern physics, Isaac Newton, was strongly motivated by religious and metaphysical questions, but he took pains to separate his theological from his scientific writings. How much more valuable and meaningful might his major works have been to us, had Newton frankly revealed his motives along with his laws and shown us the workings of his creative imagination? But at least Newton did publish his religious and philosophical thoughts. It is time now for scientists to humanize science by exposing the fears and drives within themselves and their predecessors that motivate their interest in and enjoyment of science.

There may be a culture gap between the sciences and the humanities, but there is no existential gap. Physics is relevant not only because it surrounds us with television and computers and sends us to the moon, but because it is rooted in the same human soil that nourishes all meaningful human endeavor.

THE PERSONAL APPROACH

You may be surprised to learn that this and not the previous section deals with the personal. Well, to be honest, they both do. Of necessity there is some overlap between the different approaches. In the last section I was concerned primarily with the philosophic and psychological underpinnings of physics to which the individual physicist can bear witness. Here I am concerned with simple human foibles and our groping struggles with the creative process. We strive to remove all traces of trial and effort in presenting our ideas and achievements to others. We want to exhibit a polished surface for others to admire without leaving any evidence of our struggle and toil. Of course, artists do this too. They destroy or conceal their failures, even their sketches and studies, releasing only their finished works. To some extent, this is reasonable. Who wants to look in someone's wastebasket, even Picasso's or Shakespeare's? Well, you probably wouldn't want to in place of viewing *Guernica* or watching *Hamlet*. But on the other hand, think how much we profit from seeing Michaelangelo's unfinished giants or Leonardo's notebooks.

Scientific works are austere and obscure enough for most people without concealing entirely the human labor and mistakes that went into producing them. Students need to be exposed to the struggle that scientists themselves go through in creating their theories and experiments. However embarrassing and unfamiliar it is, scientists should occasionally reveal the very human process of creation which they and other scientists go through, if students are ever to believe that scientific theories are made by real people with irrational and ridiculous thoughts, rather than by logical automatons with no human weaknesses and temptations.

As a matter of fact, it would even be useful for some research papers to take on a little of this personal flavor. Scientists could do a great service to both their colleagues and their students if they were to admit in print some of their errors and dead ends. What is really behind the conventions of scientific publication? Is there an attempt to convince others that the ideas and theories presented are really objective by eliminating from them all trace of the fallible human minds that produced them? I am not arguing against clean and elegant presentation of research work, but rather asking why this is the only allowable form it can take. Science is a human endeavor, and it surely cannot lose its value by making that fact evident.

THE INTEGRATIVE APPROACH

In the last decade, there has been a trend on college and university campuses toward interdisciplinary studies. But despite the success of fields like biophysics and psycholinguistics, and the upsurge in ecology, most scholars are suspicious of integration for its own sake. Apart from guarding of the territory of one's discipline, there is some justification for these suspicions. A field has a certain integrity of its own that ought not be tampered with arbitrarily. And students cannot branch out without first mastering the basics. There is no royal road to learning. I know from my teaching experiences that attempts to philosophize about physics without first teaching the basics are fruitless.

But part of the desire to protect physics and other fields and to keep them from becoming watered down is really a conservative tendency to keep the disciplines exactly as they are. Compromise is needed. Physics might better serve human needs by working cooperatively with art and philosophy, than by remaining an aloof and self-contained discipline. Integration for its own sake isn't really the issue. A fundamental common consideration must provide the bond. This, I think, is where some of the cross-disciplinary programs have gotten into difficulties: they haven't dug deeply enough to find the common roots of the fields being integrated.

In recent times, unfortunately, philosophy has served physics only slightly by helping to evaluate its methods and to interpret its theories and concepts. Philosophers find interesting issues in physics, but physicists largely ignore philosophy. Ever since natural philosophy broke away from the mainstream of philosophical inquiry and turned itself into modern science, it has lost sight of what the two fields shared as a vital common motive, viz., the exploration of nature in quest of the meaning and purpose of human existence. (Physicists, of course, flatter themselves that they have made their progress by rejecting such metaphysical matters, but the final judgment of history has yet to be made.) There is a deep underlying connection between the fields of physics, mathematics, philosophy (including metaphysics, epistemology, and ethics), astrology, art, music, religion, psychology, mythology, and language. This connection should be the starting point and basic theme of a more integrative course in physics. While teaching the essential subject matter of physics, its relationship to these and other fields may be explored through this basic connection which is at the same time its human heart.

THE PHYSICAL APPROACH

Lest anyone get the wrong idea, this section will emphasize the importance of the traditional physical viewpoint in teaching physics. Whatever impression I may have given to the contrary, I do not reject physics and the physical sciences. I believe we must reevaluate and cease idolizing them, but not forswear them. I have tried to show that physics *does* play a role in the pursuit of the human quest for meaning, which I think will be made most evident by recognizing the poetical, metaphorical activity that physics is. But one doesn't do that by eliminating or diluting the concerns and interests of the physicist, any more than one would learn the value of art by focusing exclusively on the representational quality of painting and disregarding its techniques.

One change in physics teaching that might prove valuable, even from the narrow scientific viewpoint, is to place greater emphasis on the structure, character, and interpretation of physical theories and concepts. This is especially true for the nonscience major. Some familiarity with problem solving and lab work is important for almost anyone studying physics, but the humanities or liberal arts student needs to get a feeling for the shape and texture of the subject and for how it makes contact with the world of experience and ideas. It is one thing to know how we use the law of gravity to predict the orbit of Mars, but it is equally important to know what this law assumes and implies about space, time, and matter, and how we view the world through it.

THE HISTORICAL APPROACH

The history of physics has also come into its own in the last decade. Many universities are developing departments of science history and offering courses in this long-neglected area. This is a good trend. But there is also a need to integrate more historical material into the garden-variety physics courses, as well.

We have a tendency to be rather smug about modern science, by which I mean science since the time of Isaac Newton and the Enlightenment. Science and the exploration of the world took a new turn with the ideas of Copernicus, Bacon, and Descartes. We began to *believe* the hypotheses that, at an earlier time, had been used only to save the appearances of things. Greek scientists and their medieval successors had looked upon geometry and astronomy as constructions devised for the purpose of reconciling the inner, transcendental truth and harmony with the outer appearances and disorder of the physical world. But starting with Copernicus and Bacon, we began to take as synonymous with truth, itself, any collection of hypotheses that explain *all* the appearances. We began modern idolatry.[6] Now our present smugness stems from the remarkable achievements that this new departure led to. But in alienating ourselves from the past, we have not only dehumanized science, but also lost sight of our integral connection with the continuing body of scientific and philosophical thought that goes back 2,500 years to the period of the pre-Socratic philosophers, and beyond that to the mythic origins of religion and philosophy. It is this vital connection that needs to be brought into the teaching of modern science.

We congratulate ourselves that the truth of today is so much more remarkable than the tales of the ancients. What we have may be more imaginative, but it is certainly no more true. We are pagans all. We look down our noses at Aristotle and his successors. How could they have been so foolish, so unempirical? They hadn't the simple sense to verify their ideas by checking them against the facts of reality. But is it any wonder that people who were more aware than we of the world as representation, as metaphor, would be less likely to look for final truth in their own constructs? Are we sure that the world would have looked to them as it does to us? How much of the modern paradigm is a product of modern consciousness? And what of astrology and alchemy? We dismiss them as pseudosciences and condemn their practitioners as having been charlatans, at worst, and fools, at best. But Jung has shown the profound significance of alchemical symbolism in understanding the mind. The great lesson that physics needs to relearn from alchemy and astrology is how to put humans back at the heart of science. We imagine the alchemists were irrational because they went through the same motions over and over for many years in the futile attempt to purify matter, to transform base lead into noble gold. But to the alchemists, their material manipulations were just an allegory of the far more important internal process of spiritual purification. Their goal was not control over nature and the raping of its wealth, but the enlightenment of themselves and the human race. How many modern scientists and engineers can say the same?

The historical view is the human view—the view of humankind's struggle and evolution. It is as important to physics as are its assumptions and motives. When we deny history, we deny ourselves.

THE AESTHETIC, IMAGINATIVE APPROACH

This approach reveals the representational, metaphorical nature of physics. It can do this in two ways: first, by exploring the aesthetic criteria of judgment and appreciation that physics shares with the arts; and second, by probing to apprehend the beauty, harmony, and meaning inherent in physical theory and its mathematical formulation.

Much work has already been done that connects physics with the arts, as, for example, explorations of the symmetry, harmony, and order that occur in the formulations of modern physics and mathematics. Symmetry plays an increasingly important role in modern treatments of elementary particles and in attempts to find some pattern behind the basic forces and concepts in physics. And, of course, symmetry is found everywhere in nature: in snowflakes, seashells, multicellular organisms, and crystal structure. The comparative analysis of symmetry and other aesthetic qualities in nature, the arts, and science will continue to be a fruitful way of sensitizing us to what they all have in common. The problem, as usual, is that such aesthetic matters are not included in ordinary physics courses but are relegated to special "cultural" presentations which are seen as interesting but hardly essential.

The second avenue of this approach, which involves the very apprehension of the beauty and meaning of physics, is much more difficult to follow in the classroom. It brings to mind a fantasy of some kind of physics appreciation class, with everyone sitting around "digging" quantum theory. (Maybe this isn't as ridiculous as it sounds.) Part of what I mean by apprehension could be accomplished by having students read the testimony of great scientists on the inspiring quality of physical theories: such works as Einstein's *Autobiography* and Werner Heisenberg's *Physics and Beyond*. But as valuable as such books are, they still present us with vicarious experience, seen through the eyes of another. What I am referring to is something more direct, something more like the musical experience I attempted to evoke with the discussion of Cantor's treatment of the infinite. There, the very form and structure of the arguments—their ability to build up to some kind of a climax or finale and to reach beyond themselves and call forth transcendent meaning—can give us a direct sense of awe and appreciation akin to our experience of art.

It is this evocative quality of physics and mathematics that we need to bring directly into the classroom. It is one of the most elusive things to define, let alone to capture and communicate to others. But it is at the heart of physics as metaphor. And it is essential to the full appreciation of physics. It comes close to the whole matter of meaning and value, as does the idea of metaphor. If science and the physical world are humanly created symbols and meanings, then whatever causes us to stand dumbfounded before a waterfall or a shooting star is the same thing that inspires us when we contemplate an equation or a painting. Meaning is contained in all these things. They are metaphors, representations, not of external objects and ideas, but of some deep inner vital force, of the inchoate meaning of life itself. They embody the collective, racial understanding of the purpose and value of our existence, as we have gleaned it so far. They are as vital and nourishing to us as wholesome food.

To attempt to find effective illustrations of this metaphorical, evocative quality in physics is very difficult. In part, it is so personal a matter that it seems hopeless to try transmitting it to others. But it is precisely because the subjective can be objectified that we have collective representations. If we wish to reclaim our human projections, we must begin by exposing what is human in them. Furthermore, the contemplative conjuring a teacher of physics must go through to locate the vital and aesthetic quality of some physical theory for a class presentation is itself a profound experience. It requires a kind of imaginative exercise in which we become self-consciously aware of what we *do* to project, apprehend, and appreciate the world. It requires an enhancement of the figuration process I spoke of earlier, which is the conversion of sensation into concept and image, as, for example, when we hear a thrush singing or when a blind person sees a room through his or her cane. As we study and reflect on ideas and theories in physics, we unconsciously or imperceptibly construct images and pictures which transcend the words and mathematical symbols that express them. These images have meaning and value which cannot be said to reside independently in the symbols, but which the symbols carry and transmit to us as metaphorical agents of other minds. In contemplating these abstract theories and ideas, the created meaning of others comes to life in our own minds. And it is this coming to life, this element of the active imagination, which is almost completely absent from the teaching of science.

Some years ago I was teaching an introductory physics course for liberal arts students and elementary education majors. I was searching for a way to capture the feeling of the continuity between early peoples' sense of wonder about the cosmos and the present-day study of physics. I had decided to use the image of ancient shepherds whose nocturnal contemplation of the heavens could be seen to foreshadow the modern astronomer's midnight star watch. But I wanted to do something more than just talk about it. I sought in vain for some visual aid that would create a convincing starry night for all to survey, as it were, with the ancient shepherds. No star projector or slide machine seemed capable of providing the proper sense of illusion. The thought of using some rickety light-leaking device in the fake partial darkness of a college lecture hall to evoke the splendor of the ancient night left me completely cold. Finally I hit upon a bold solution, and summoning up my courage to go through with it, I entered the lecture room with my heart in my mouth. I set the stage as poetically as I dared, trying to call up the quality of life of the ancient star gazers and the wondrous heavenly spectacle that filled their peaceful nightly wanderings. And as I asked my students to gaze toward the ceiling and imagine what the shepherd saw and felt, I simply extinguished the room lights and fell silent.

After a full minute of darkness and silence, I tried to prod a few sluggish imaginations by quietly suggesting closing the eyes and looking for the Milky Way, Orion, Betelguese, Sirius, and the Great Bear. It was with some reluctance, after another minute or two, and probably with a rather red race that I turned the lights back on. And although there were a few people who seemed puzzled or embarrassed, there were other faces that were smiling happily and appreciatively and telling me what I wanted to know. Later one student said to me that although she hadn't visualized very much, just having a few minutes of darkness and silence in the middle of a lecture had forced her to stop taking notes and concentrating so hard and had given her a rare opportunity to relax and to contemplate a side of physics she had never thought about before.

Since that time, I have used other evocative techniques in different classes, but never one quite as open-ended and with as much potential for exercising the creative imagination. Almost anything in physics that appeals to an instructor can be used as a starting point. The sky's the limit (or rather no limit, as I discovered). One of my own favorites is the theory of relativity with its marvelously provactive curved spacetime, geometrized gravity, and time paradoxes.

Incidentally, this is a good place to acknowledge the debt I owe to Louis Pauwels and Jacques Bergier. Their book, *The Morning of the Magicians,* first made me explicitly aware of the evocative, analogical quality of physics, so useful in teaching it humanistically. In a brilliant chapter called "The Magic Mind Rediscovered," Pauwels and Bergier contrast the two major aspects of mind: the *digital* (logical, rational, sequential) and *analogue* (intuitive, imaginative, creative). Exploring both aspects in science, art, and religion, they view today's physical theory as a modern *argot* or slang, embodying in analogue form, the mysteries and secrets of the universe, just as the Gothic cathedrals of the Middle Ages did through architecture and sculpture.

Exercises, Puzzles, Speculations

In this final section, I shall suggest some matters to contemplate and think over which might be evocative in the way I just spoke of. This material may provoke the imagination or stimulate further thought in directions suggested earlier in the book. The ideas are, for the most part, tentative and speculative. This is intentional. For I am not attempting to sew up all the loose ends in the fabric of physical metaphors. Rather, I shall try to contribute a few swatches and patterns to a patchwork quilt, on which many people are working together and whose real comfort and warmth is not yet apparent to any of us.

PERCEPTION WITHOUT INTERPRETATION

Walk down a street. What do you see? Houses, trees, cars, people—all in a passing evolving form that we tend to interpret in terms of motion and perspective. We know, for instance, that we are approaching and passing a tree because we see its trunk enlarge and because we see a varying, rotating view of its bark as well. If we look back after passing the tree, we see a rotated image of its other side, as if the tree had turned through 180° as we passed it, and as we move on, we note the image diminish in size. Of course, we do not associate the enlarging, rotating, and diminishing of the tree image with the tree itself, but rather with our motion relative to the tree and the laws of perspective which govern the sight of objects in depth and at distances.

In fact we don't usually think of what we see as an *image* at all, but simply as a tree. We are, for the most part, unconscious that light is entering our eyes and forming a pattern on the retina, and that this pattern of visual sensations must be interpreted by the mind as the articulated objects we see. Similarly, blind people are largely unconscious of the cane and series of jolts on their palms through which they *see* their environment. Most people easily recognize the profound difference between the actual touch-sensations of blind people and the objects of the physical world: they sense the process of figuration in their minds. Yet we fail to recognize the equally profound difference between our own visual sensations and the objects projected from them. We are unconscious of our own figuration. We are so accustomed to the conventions of sight by which we infer the existence and motion of objects in space that it requires a very deliberate, concentrated act of the will to sever, even conceptually, the raw data of our sensations from the constructed interpretations of that data.

The purpose of this exercise is to take a first step toward becoming aware of the mind's figurative creative process by consciously trying to stop it. The next time you walk down a street (preferably a fairly safe, untrafficked one), try to be aware of the pure sensations you experience, and, at the same time, try to avoid interpreting them. That's not a tree trunk you're passing; it's an elongated vertical brown area in your visual field. It enlarges, its surface texture changes and then it grows smaller. It isn't all *due* to your perspective view and relative motion. It's simply a raw brownness, enlarging, changing, diminishing, nothing more. That's not a car passing by; it's not a discrete object at all, just a varying portion of the general visual background. That's not sky, treetops, and roofs above your head, but simply a new set of colors and patterns that your visual field has taken on to the accompaniment of certain kinesthetic feelings in your head and neck (as you tilt your head back). What you *see* is just a variable continuum of shapes and colors: no delineated things separated from each other by space, but a multicolored unified pattern of variations. (Nearsighted people can make this exercise graphic by simply removing their eyeglasses.) There is no far and near, above and below, left, right, before, and behind. There is no depth or distance; only that raw visual experience which isn't taking place in *space,* at all, but in your eyes, in your head, in your mind. (And where, pray tell, is that?)

I don't recommend this exercise as a steady diet (any more than I'd recommend rejecting our metaphors permanently), but only as an experiment in becoming aware of figuration, of the way we participate in creating our environment and of the relativity of our experience of the world. Many variations on this theme will suggest themselves. The blind walk used in encounter and sensitivity groups is one. You are blindfolded and led by someone through a series of sound, taste, touch, and smell experiences. This can be a marvelous adventure. But if you should try it as an experiment in figuration, it is important not to dwell only on the heightened feelings. It is not just sensation you want to enhance, but imagination. Try to remain contemplative and reflective about what happens to you. Muse over what you do with your mind and how you contribute to the construction of what you take to be out there. It is a combination of intellectual/critical mental activity with heightened creativity and imagination that we are after. If we can learn to intensify and illuminate the act of participation, we shall come to understand just how profoundly creative all human beings really are.

SELF EFFECTS AND THE RUSSELL PARADOX

It always strikes me as curious that two very perplexing problems of twentieth-century physics and mathematics are related to each other philosophically and that, when viewed together, they shed light on the illusion of the subject/object dichotomy. These are the paradoxes of self effects of elementary particles in physics and of the concept of the set of all sets in mathematical logic. They seem to me intimately connected with our presumption of a multiple world, which, sooner or later, leads to the paradox of trying to separate a thing from itself, or (to use Alan Watts's provocative example) of an eye trying to see itself.

The matter of self effects has always plagued physicists. It occurs, for example, as an inevitable contradiction between the ideas of an elementary particle and the force field it manifests. How, for example, does the electric field of an electron affect the electron itself? Why doesn't the electron explode under the action of its own repulsive forces? If an electron is made up of some kind of distributed negative electric charge, why doesn't the mutual repulsion of all this concentrated negative charge force the electron to fly apart, just as two electrons that are brought together and released will fly apart? Is the electron immune to its own force? How is this possible? The laws of physics imply that all electric charge is susceptible to the electric field. Is there some stronger force that holds the electron together? But what is the nature of this force? Wouldn't this new force also be generated by some elementary particle as its source, just as the electron is the source of the electric field? And what would keep this new particle from exploding or collapsing under its own influence? Or are we to introduce a sequence of stronger and stronger forces ad infinitum, each successive force explaining its predecessor?

Perhaps we can get around the whole problem by assuming that an electron is just a point with no finite size or spatial extension. But the electron has a finite mass and electric charge. It would then have an infinite density since the mass and charge would occupy zero volume. And how can a physical entity be a point? Matter in physics always takes up space. If a point electron were assumed to exist, would it be subject to physical laws different from those that govern the external behavior of electrons?

Modern physics gets around this kind of problem essentially by defining it out of existence. The theoretical picture of the electron, which we have seen does not explain how the electron can exist and be stable, is redefined or *renormalized* as the observed physical electron. The justification is the usual one that it works. We saw earlier that this pragmatic approach can be a kind of self-fulfilling prophecy. But, in this case, renormalization is not very satisfactory on any philosophical grounds.

Let's now turn to the problem in mathematical logic, which first upset Cantor's apple cart and eventually led to the discovery by Gödel that all the apples have worms in them anyway. It was by using Cantor's set theory that mathematicians hoped to axiomatize all of mathematics. As we already saw, this was finally proved impossible by Kurt Gödel. But the first ringing alarm was sounded by Bertrand Russell with his famous paradox of the set of all sets. A simple illustration of Russell's paradox is the following: All books may be placed in one of two categories: A—those books that mention or refer to themselves, and B—those that do not. Most novels, for example, are in category B, since few writers of fiction have occasion to speak about their books in the course of writing them. Most nonfiction, on the other hand, is in category A, since prefaces and introductions in works of nonfiction often refer to the book at hand. Clearly every book is either of type A or of type B without exception. But what of a special reference volume that lists *all* books of type B, those that do not refer to themselves, and no other books? Is this reference book of type A or B? A little reflection shows that neither alternative is logically possible! Suppose that the reference does list itself. Then it is of type A because it refers to itself. But since the reference lists only books of type B, and is itself of type A, it must *not* list itself. All right then, we must assume that it does not list itself. That means it is of type B, since it doesn't mention itself. But it is supposed to include all books of type B and therefore must list itself. (There is no trick way out of this.)

Logicians have gotten around this problem by introducing a hierarchy of categories: a set of sets is not itself a simple set, but a kind of superset, not subject to the laws of sets. This, I think, is just as arbitrary as renormalization in physics. There is nothing in the basic definition of a set, as a collection of anything however abstract or concrete, that would exclude sets themselves from being elements of sets.

Both these paradoxes, the physical and the mathematical, hint at the fundamental error of breaking things up and treating them separately. They reveal the illusory nature of our metaphors that assume separate individual entities exist in an extended space. The concept of an electron as used in physics may be a contradiction in terms, so long as an electron is conceived of and treated as distinct from all other electrons and other things in the universe.

The categorizing and naming of different objects, which underlies the paradoxes of set theory, also conceals the basic assumption that the world consists of distinct, divisible things. Our very thought processes and language reflect the illusion of separateness and disunity. The fundamental analytic method of science is at stake here. Can cause and effect, mind and matter, subject and object ever really be treated as separate and discrete? For the resolution (or is it dissolution?) of these dilemmas, I look to the renunciation of idolatry and the unification of science and the humanities.

THE ELECTRON-MUON PUZZLE

I want to consider one further problem in physics because it is so suggestive of some of the ideas in this book—ideas involving Cantor's cardinality of the continuum, and the fact that the size or measure of a space bears no relationship to how many points it contains. What this has to do with elementary particles in physics will, I hope, become clear shortly and provide some food for thought.

Most people are already familiar with the electron as the tiny negatively charged constitutent of an atom and the carrier of electrical current in metals. But what, you may well ask, is a muon? Well, physicists themselves don't really know, and that's part of the problem. We can, of course, describe the muon's physical properties: It is an elementary particle about 200 times as massive as an electron. It may be either negatively or positively charged, just like the electron and its positively charged antiparticle, the positron. It does not last indefinitely like the electron; its average lifetime is about a millionth of a second (actually a rather long lifetime for an elementary particle), but it has the same intrinsic *spin* that an electron has. In fact, all of its other known properties are the same as those of an electron. Apart from the fact that muons are more massive than electrons and are not stable (i.e., not immortal), they are identical to electrons in all other respects.

You may wonder what is unusual about this. After all, many other particles have some of their properties in common, and this even provides the basis for classifying them. But with all other particles, their differences are significant in determining the roles they seem to play in the scheme of things. Neutrons, for instance, are slightly more massive than protons, but, unlike positively charged protons, neutrons are electrically neutral, and this is crucial in providing the stability of the atomic nucleus, which is made up of neutrons and protons. The mass of the pion is directly related to the range of the nuclear force. And the different charge states of the pion (positive, negative, and neutral) explain the nuclear force of attraction between particles, which are themselves differently charged (a positive pion accounts for the strong nuclear force between a neutron and a proton, whereas a neutral pion accounts for that between two protons or two neutrons). The point is that while we are far from understanding the proliferation and variety of elementary particles, we can generally rationalize the existence of each, in terms of its unique characteristics. The function of each elementary particle is intimately related to its properties. At least until we get to the muon.

The muon, as far as we can tell, plays no essential role in the cosmic scheme and appears to be a kind of fat, short-lived electron. A great many clever and complex experiments have been devised to detect some significant difference between the electron and the muon, other than their differences in mass and lifetime. All such experiments have found a null effect, that is to say, that the electron and the muon behave exactly alike in all their interactions with matter. There is no difference between them that is of any significance in nature. It would seem that the universe can get along perfectly well without muons. (They do figure in the disintegration of pions and other particles, but there are other modes of decay available in all cases.) In the cosmic labor market, it seems that electrons are heavily overworked. But apparently electrons are very efficient and don't require any assistance. So why does the muon exist when it can do no more than what an electron already does and when, in fact, it is never even called upon to pinch-hit?

This problem has intrigued me for many years, and, without any pretense to an answer, I shall offer a few speculations. What fascinates me is that the electron and the muon appear to differ in only one significant respect, apart from their unequal lifetimes. The muon is roughly 200 times more massive than the electron.

In other words, they differ in one important quantitative measure —mass. Why should we have two different versions of what appears otherwise to be the same particle? Perhaps we need to ask a different question. In what sense can we say that the electron and the muon are different things? What else can appear to have different measures or sizes and yet retain its essential individual character? Well, what comes to my mind is the line segment in Cantor's famous proofs that we discussed in chapter 6. There we saw that all line segments are equivalent in cardinality regardless of their lengths. Or more simply, the "number" of points in all line segments is the same: a one-to-one correspondence may be established between the points in any two line segments. This basic equivalence holds in two and three dimensions as well. Thus all surface areas and spatial volumes, despite the differences in their measures (area or volume), are equivalent when considered as sets of geometrical points. Two spheres of different volume have the same number of points and in that sense may be considered equivalent.

But what has this discussion about points and geometrical spheres to do with physical particles? Well, remember I said this was going to be speculative. I am going to assume that electrons are made out of mathematical points. Recall that physics as metaphor means not taking the physical too literally. Besides, if elementary particles are not geometrical entities consisting of abstract points, what is it that they are made of? Modern physical matter is not very substantial stuff, and quantum theory gives us little hope of anything other than an abstract mathematical basis of things.

In any case, I shall assume that somehow electrons and muons are made of points and that their mass is distributed in their volume. They need not be hard little balls. They certainly need not be impenetrable, but we shall imagine them as some kind of continuous, if tenuous, spheroids. Their mass or volume is simply a measure of their size, which isn't very significant if we take Cantor's analysis seriously. Other particles differ by several significant measures, but the electron and the muon are like two line segments, made in the same way out of the elemental points but having different lengths. The electron and muon are balls of different size, but with the same number of points. Well, how significant is this one measure, this volume or mass? What is it that makes two line segments differ by length? Can the points be said to take up more space in one case than in the other? What is it that makes lines, surfaces, and space regions have different lengths, areas, and volumes in the first place? Here is a great mystery which has no answer. Probably it hints, once again, at the illusory nature of our space metaphor with its fixation on the notion of extended space. If space and its extension are imaginative projections of the mind, then stretching and shrinking them shouldn't seem so impossible. Size itself, measure, and number are mere appearance and metaphor, and they should not be mistaken for some ultimate invariant—they should not be idolized.

But in the case of the three-dimensional measure, volume, there is yet another consideration that must give us pause. This is the astounding and paradoxical theorem of Banach and Tarski[7] which says that a sphere of any given size may be taken apart and reassembled into a sphere of a different size. Furthermore, this may be done in a *finite* number of steps and *without changing the size or shape* of the disassembled pieces. In other words, a pea may be taken apart, like a three-dimensional jigsaw puzzle and put back together to form the sun! And an electron may be transformed into a muon in a finite number of steps (in one version of the proof, five). Such transformations the mind may know how to perform far better than we. This incredible theorem of Banach and Tarski makes us think twice about that cardinal concept of measure and quantification by which we all set such great store. To the extent that matter today is represented by some kind of an abstract distribution in a mathematical space, to that extent all our measures are, in some sense, subject to the arbitrary whimsy of the Banach-Tarski transformation. What is it that keeps everything under control and in place? What is it that sustains our metaphors? What is size? "There's safety in numbers," we always say. But is there?

CHILDHOOD SPACES

In recent years, through the brilliant pioneering work of Jean Piaget and his followers, we have learned a great deal about the perceptions of children. I have been especially interested in the work done on children's perception of space and the formation of the object concept. I should like to point out that in our interpretation of childhood perception we may well be guilty of space idolatry.

I shall illustrate this with the game of peek-a-boo that babies love. Most little children are delighted by the sight of a head bobbing up and down at the edge of a crib. They are overjoyed at seeing someone enter and leave their field of vision. Piaget helps us interpret this behavior. He tells us that at this early stage children have not yet learned to *conserve* objects. This means that in the conceptual world of children, there is no clear sense of the continuity and persistence of objects in space and time. They are surprised by each appearance of an object because they have no reason to expect its continued existence once it has disappeared from their view. They have not yet formulated an adult conception of space and time in which they can interpret depth and breadth and picture objects in memory as they move out of sight behind other objects. They must depend entirely on their visual field for information and have not yet learned to use past experience and memory to draw inferences about their world. They are very close to practicing the perception without interpretation that was discussed in the opening exercise of this section.

Now why have I suggested that this description of a child's space perception may be idolatrous? Because in thinking about all of this we usually assume that the child lives in and experiences the same space that we do. By the same token, we are in the habit of imagining that primitive people lived in the very same physical world that we live in and that it was only ignorance and superstition that kept them from understanding and interpreting it as well as we do. This intellectual stance ignores the evolution of consciousness and the impossibility of distinguishing changes in some presumed external environment from changes in human consciousness itself. A child's mental evolution may well be a case of ontogeny recapitulating phylogeny—the evolution of an individual's consciousness repeating and telescoping that of the race. Just as R. D. Laing suggests that we should learn from the insane rather than only try to cure them, so it may be that we ought to try learning from our children and not rush them toward an adult view of the world.

What may be delighting little children is their witnessing the acts of creation and destruction as they see us appear and disappear before them. These experiences in which they participate with us are as natural to their world as endurance and continuity are to ours. They accept them without interpretation, but they delight in the novelty of creation as we all do. We look retrospectively at the space of the child and think of it as lacking in perspective, continuity, and individuality and as reflecting the embryonic experience of the womb. But what we are really talking about is a more organic, unified, chaotic sense of space. It is not a space lacking or deficient in anything, but rather one that is different from ours—another metaphor. The evolution of the infant space metaphor into our adult one deserves deep study and contemplation, for it can reveal some of the secrets of existence and creation. Childhood space, viewed without idolatry, can open vistas to us.

THE MYSTERY OF BIRD NAVIGATION

As a final example of a mind-stretching exercise or puzzle, I want to speculate about bird navigation. Researchers have discovered that birds can orient themselves in space in several different ways. They use the position of the sun in conjunction with a kind of internal bioclock; they take advantage of prevailing winds and weather fronts; they sense the magnetic field of the earth; they employ a crude sense of smell to recognize home; and they steer a course by the stars, identifying the pole star as the one that does not rotate. All of this was discovered through many interlocking and carefully controlled experiments in which birds are forced to depend on only one of these mechanisms to navigate or, in the case of homing pigeons, to return to their home destination. Some of these experiments are ingenious. For example, to isolate the magnetic sense, scientists temporarily impaired the vision of a group of birds by placing frosted contact lenses over their eyes, and then had them fly in heavily overcast weather. The birds could use neither landmarks nor the sun. The magnetic navigation system of some of the birds was then incapacitated by attaching magnets to their bodies or placing demagnetizing coils around their heads. In this way, without the sun to guide them and with no visual acuity to speak of, the magnetic nature of birds' directional sense was established, since the visually impaired birds navigated significantly better than those that were both visually and magnetically impaired.

In another series of experiments, the flying pattern of many birds was observed inside a planetarium that projected a normal image of the stars rotating around the pole star. A certain direction in the flying pattern of the birds predominated statistically in this normal case. But when the stars were not rotated, the bird pattern became random, and when a different star was made to be the pole by rotating the sky around it, the predominant direction of the birds changed correspondingly. These and other experiments have given us an amazing picture of the means birds use to navigate. The researchers readily admit, however, that their understanding of these steering mechanisms is extremely crude and that they are still largely puzzled by how birds manage, with remarkable accuracy, to pinpoint a destination many thousands of miles away. And the deepest mystery of all is how the birds know where they are supposed to go, in the first place.

I cannot help but wonder what this research would look like if approached from the viewpoint of alternative space metaphors instead of our common idolatrous one. We try to understand bird navigation in terms of spatial and causal mechanisms. All these explanatory mechanisms seem to place the cart before the horse. Star, sun, and magnetic navigation represent abstract projections of subjectively experienced relationships. Earlier forms of consciousness experienced these as organic, astrological, or mythic. It is from our internally felt sensations and their relationships that we build our notions of space and project an external world. Is it any wonder now that we should find gravity, magnetism, and sunlight *out there*?

How else are we to understand birds flying great distances "through space" and doing pinpoint "navigation?" Perhaps for the bird, it's more like sitting in a movie theater and watching the flight go by on the screen, knowing all the while that the destination is a fore-ordained and recorded there at the end of

Sitting in a movie theater and watching the flight go by on the screen.

the film. The bird must supply the creative energy, but it's more like running the projector than physically moving through space. The whole thing is coiled up on the film in the bird's head. He doesn't go anywhere; he just runs his movie. How can he miss a target? It was there on the film all along. It's all very much like the business of numbers and time in chapter 6. There's the *coiled up* state in which numbers and time are symbolic, synchronistic, potential, and simultaneous. And there's the *uncoiled* state characterized as sequential, causal, actualized, and linear. What appears as a causal sequence to us may be more of a synchronistic, mythic, dynamic pattern to the bird. For him it's like von Franz's fluttering dragonfly. He's got all his necessary information coiled up inside of him, maybe imprinted somewhere in his cells along with the genetic code. There must be some reason why we picture the DNA molecule as a spiral.

You protest. How could anyone ever know such things? How could we ever have faith in such information or test it? Well, we might learn to trust inner voices, or to hear a little bird whispering in our ears—a cardinal, perhaps.

Notes

Notes

Introduction

1. C. M. Turbayne, *The Myth of Metaphor* (Columbia, S. C.: University of South Carolina Press, 1970), p. 22.

2. F. Capra, "Quark Physics without Quarks: A Review of Recent Developments in S-Matrix Theory," *American Journal of Physics* 47, no. 1 (January 1979), p. 12.

Chapter 1. The Lengths We Go To

1. If the meter stick is shorter than the table, I must mark off the length of the meter stick along the edge of the table a number of times. Making these marks once again requires decisions about the coincidence of two points, adding nothing to our analysis.

2. We might instead decide which mark on the meter stick the right end of the table is closest to. But this, like the decision about the coincidence of two points, is an estimation of distance and is subject to the same criticism of circularity and subjectivity.

Chapter 2. The Stacked Deck

1. H. Poincaré, *The Foundations of Science* (New York and Garrison, N. Y.: Science Press, 1913); M. Polanyi, *Personal Knowledge* (New York and Evanston: Harper and Row, 1964) and *The Tacit Dimension* (Garden City, N.Y.: Doubleday, 1967); T. S. Kuhn, *The Structure of Scientific Revolutions* (Chicago: University of Chicago Press, 1962); G. Holton, *Thematic Origins of Scientific Thought* (Cambridge, Mass.: Harvard University Press, 1973); H. I. Brown, *Perception, Theory, and Commitment* (Chicago and London: University of Chicago Press, 1977).

2. Incidentally, it turns out in relativity theory that a hyperbolic function of distance divided by time is the most convenient measure of speed. It obeys a law of addition of velocities which our ordinary measure of speed does not.

3. M. Polanyi, *Personal Knowledge*, p. 16.

4. R. M. Pirsig, *Zen and the Art of Motorcycle Maintenance* (New York: William Morrow, 1974), p. 268.

5. That decisions about the degree of order in a pattern and the likelihood of its occurrence are matters of personal appraisal is argued effectively by M. Polanyi in *Personal Knowledge*. See especially chapters 2 and 3.

6. See A. S. Eddington, *The Philosophy of Physical Science* (New York: Macmillan, 1939), p. 123.

Chapter 3. Space to Let

1. O. Barfield, *Saving the Appearances* (London: Faber and Faber, 1957), p. 78.

2. See, e.g., O. Barfield, *Saving the Appearances*; E. Grant, *Physical Science in the Middle Ages* (Cambridge: Cambridge University Press, 1977).

3. If finite, then inconceivably large.

4. The big bang theory makes much of the fact that the background, ambient temperature of empty space is just under three Kelvin degrees above absolute zero, not a very comforting temperature.

5. T. S. Kuhn, *The Structure of Scientific Revolutions*.

6. C. Jung, Introduction to *The I Ching or Book of Changes*, the Richard Wilhelm translation rendered into English by Cary F. Baynes, Bollingen Series XIX (Princeton, N. J.: Princeton University Press, 1967).

7. O. Barfield, *Saving the Appearances*, p. 103.

8. *Ibid*., p. 104.

9. P. D. Ouspensky, *Tertium Organum* (New York: Knopf, 1959), p. 106.

10. W. Follett, *Modern American Usage* (New York: Hill and Wang, 1966), p. 257.

Chapter 4. Time for a Change

1. See P. D. Ouspensky, *Tertium Organum*.

2. I use the verb *qualify* in the sense of attribute quality to.

3. B. L. Whorf, *Language, Thought, and Reality* (Cambridge, Mass.: M.I.T. Press, 1956), p. 151. Excerpts reprinted by permission of M.I.T. Press.

4. *Ibid*.

5. Sri Krishna Prem, *The Yoga of the Bhagavat Gita* (Baltimore: Penguin, 1973), p. 207.

6. *Ibid*., p. 209.

7. *Ibid*.

8. T. Dantzig, *Number—the Language of Science* (New York: Macmillan, 1946), p. 168.

9. O. Barfield, *Saving the Appearances*, p. 64.

Chapter 5. What's the Matter?

1. B. Russell, *The Analysis of Matter* (New York: Dover, 1954), p. 385.

2. N. R. Hanson, *Patterns of Discovery* (Cambridge: Cambridge University Press, 1972), p. 119.

3. O. Barfield, *Saving the Appearances*, p. 69.

4. *The I Ching*, p. 68. Excerpt reprinted by permission of Princeton University Press.

Chapter 6. The Numbers Racket

1. M.-L. von Franz, *Number and Time* (Evanston: Northwestern University Press, 1974), p. 45.

2. *Ibid.*
3. *Ibid.*, p. 115.
4. *Ibid.*, p. 49.
5. T. Dantzig, *Number – the Language of Science*, p. 9.
6. M.-L. von Franz, *Number and Time*, p. 237.
7. *Ibid.*, p. 157.
8. A. Daniélou, *Hindu Polytheism*, Bollingen Series LXXIII (Princeton, N. J.: Princeton University Press, 1964 and London: Routledge & Kegan Paul, 1964), p. 263. Excerpts reprinted by permission of Princeton University Press and Routledge & Kegan Paul.
9. *Ibid.*, pp. 263, 264.
10. David White has pointed out to me that the root of the Hindu word *maya* is *ma*, which is cognate with our English word *measure*!
11. G. F. Parker, *A Short Account of Greek Philosophy* (New York: Harper and Row, 1967), chapter IV.
12. *Ibid.*, p. 28.
13. In modern atonal music, there are still simple mathematical relationships underlying the twelve-tone scale.

There is an avant-garde trend in some contemporary music to throw melody and harmony out altogether, and to define as music whatever combination of noise and silence a composer concocts. This, it might be argued, is an example of inharmonious or disordered music, but I should prefer to think of it as the musical metaphor forcibly reduced to chaos.

14. E. T. Bell, *Men of Mathematics* (New York: Simon and Schuster, 1965), chapter 29, p. 555.
15. See E. Nagel and J. R. Newman, *Gödel's Proof* (New York: New York University Press, 1968).
16. Note that π is not rational even though it is defined in terms of a ratio. π is not a ratio of numbers but of lengths. If either the circumference or the diameter of a circle has a length that may be represented by a rational number, then the other will not.
17. For a full, but eminently understandable presentation, see chapter 2 of E. Kasner and J. R. Newman, *Mathematics and the Imagination* (New York: Simon and Schuster, 1956). This book, incidentally, is a wonderful introduction to the beauties of mathematics.
18. See R. J. Cohen and R. Hersh, "Non-Cantorian Set Theory," *Scientific American* 217, no. 6 (December 1967), p. 104.

Chapter 7. Death and Transfiguration

1. S. Freud, *Beyond the Pleasure Principle* (London: Hogarth Press and Institute of Psychoanalysis, 1961); B. Brophy, *Black Ship to Hell* (New York: Harcourt, Brace and World, 1962); N. O. Brown, *Life against Death* (New York: Random House, 1959).
2. E. Becker, *The Denial of Death* (New York: Free Press, Macmillan, 1973).
3. For further insight into this matter, see the interview with Becker in *Psychology Today*, April 1974.
4. E. Becker, *The Denial of Death*, p. 26. All excerpts from this work are reprinted with the permission of Macmillan Publishing Co., Inc.
5. *Ibid.*, pp. 4, 5.
6. *Ibid.*, p. 56.
7. *Ibid.*, pp. 218, 219.
8. *Ibid.*, p. 90.

9. *Ibid.*, p. 91.
10. *Ibid.*, pp. 198, 199.
11. *Ibid.*, p. 258.
12. *Ibid.*, p. 282.
13. *Ibid.*, p. 279.
14. *Ibid.*

Chapter 8. Participation without Representation

1. J. G. Neihardt, *Black Elk Speaks* (New York: Pocket Books, 1972).
2. See *Re-Vision* 1, no. 3/4 (summer/fall 1978).
3. O. Barfield, *Saving the Appearances*, p. 20.
4. M. Polanyi, *The Tacit Dimension*, pp. 12, 13.
5. O. Barfield, *Saving the Appearances*, p. 161.

Chapter 9. New Metaphors for Old

1. H. Poincaré, *Foundations of Science.*
2. M. Polanyi, *The Tacit Dimension.*
3. T. S. Kuhn, *The Structure of Scientific Revolutions.*
4. G. Holton, *Thematic Origins of Scientific Thought.*
5. H. I. Brown, *Perception, Theory, and Commitment.*
6. See O. Barfield, *Saving the Appearances*, and E. Grant, *Physical Science in the Middle Ages.*
7. See N. J. Fine, *An Introduction to Modern Mathematics* (Chicago: Rand McNally, 1965), p. 292. Also S. Banach and A. Tarski, *Fundamenta Mathematicae* 6 (1924), p. 244, and R. M. Robinson, *Fundamenta Mathematicae* 34 (1947), p. 246.

Suggestions for Further Reading

Suggestions for Further Reading

Barfield, Owen. *Saving the Appearances.*
London: Faber and Faber, 1957.

My debt to this seminal book cannot be overstated. First published in 1957, it brought forcefully to light both the essential role of consciousness and its evolution in the history of science. Human experience of the world has evolved from an ancient organic participation with nature to our modern objectified separation from it. Since experiences cannot be divorced from consciousness, our perceptions amount to mental constructs or representations. When we treat these representations as reality, forgetting that they were intended only to "save appearances," we commit a profound idolatry. In this and other writings, Barfield soundly attacks the destructive, modern belief in a cosmos distinct from and independent of consciousness.

Becker, Ernest. *The Denial of Death.*
New York: Free Press, Macmillan, 1973.

A brilliant synthesis of Freudian and existential thought. Becker's theme of human activity as a self-justifying attempt to give meaning to life and to deny death is the seed idea behind my discussion of how the fear of death has motivated and determined our modern cardinal metaphors. A powerful struggle between our need for a pacifying salvation, on the one hand, and the terrifying truth, on the other, pervades and vitalizes this book.

Blythe, R. H. *Zen in English Literature and Oriental Classics.* New York: Dutton, 1960.

My favorite book on Zen, it approaches doing Zen rather than reading about it. Artfully discussing stories and poetry from both the East and West, Blythe makes vivid for the reader the experiences of Zen and the impossibility of a strictly rational-verbal approach to life. The chapter on Don Quixote as a Zen master is brilliant and delightful.

Capra, Fritjof. *The Tao of Physics.*
Berkeley: Shambala, 1975.

One of several excellent recent books that explore the commonality between modern physical thought and mystical wisdom. Capra concentrates on the East and gives us many beautiful Hindu, Tantric, Taoist, and Zen myths and concepts which presage ideas in quantum theory and relativity. The book suffers slightly, in my opinion, from a subtly superior attitude, favoring the Western scientific approach, as if to say, "Now that we have the vindication of the physicist, it's okay to believe in the mystic."

Einstein, Albert, and **Leopold Infeld**. *The Evolution of Physics.* New York: Simon and Schuster, 1954.

One of the best popular introductions to both classical and modern physics. It does a fine job on such pivotal concepts as fields and energy, while maintaining Einstein's inimitable folksy style. As an added bonus, there is the master's down-to-earth explanation of relativity and a rare, full treatment of his own interpretation of quantum mechanics.

Eiseley, Loren. *The Immense Journey.* New York: Random House, 1957.

All of Eiseley's books are well worth reading, both for the information they contain and for his moving humanization of science. In this poetic exploration of nature and our relation to it, Eiseley distills life's endless mysteries from his own experiences. His wandering meditations on the past and present are effortlessly woven into the story of humanity.

Heisenberg, Werner. *Physics and Beyond.* New York: Harper and Row, 1972.

An autobiographical account of the origins of modern physics by one if its principal creators. The book is notable not only for its homespun presentation of the central ideas in twentieth-century physics, but also for the lively story of the battles among the giants of physics and their clashing philosophies. Heisenberg treats us to a rare personal and inside view of how science is made, and of the very human motivations, dreams, and fears that go into the process.

Holton, Gerald. *Thematic Origins of Scientific Thought.* Cambridge, Mass.: Harvard University Press, 1973.

Holton's thesis—closely related to Kuhn's (see below)—is that the central ideas behind many scientific theories are often subjective and nonrational in motivation. The fine account of Kepler's long struggle against noncircular orbits for the planets, in an attempt to preserve the central role assigned to the sun by the Greeks, is alone worth the price of the book. There is also a careful and illuminating essay on Einstein's original thought and motivation, while he was developing the ideas of the special theory of relativity.

Kasner, Edward, and **James R. Newman**. *Mathematics and the Imagination.* New York: Simon and Schuster, 1956.

The discovery of this book in high school turned my early interest in mathematics into a full-blown love affair. I still find it the best introduction to the concepts and delights of mathematics. The chapter on Cantor is a gem. It's a credit not only to Kasner and Newman but to Cantor himself that the incredible and profound theorems on infinity can be understood and appreciated at this level.

Koestler, Arthur. *The Roots of Coincidence.* New York: Random House, 1972.

A far-ranging essay attempting to reconcile ESP phenomena through the acausal, nonmaterialistic conceptions of modern physics. The thoughtful discussions of probability, causality, synchronicity, and holism, and of the theories and ideas of quantum physics are lucid and provocative. Koestler's "rapprochement between the conceptual world of parapsychology and that of modern physics is an important step towards the demolition of the greatest superstition of our age—the materialistic clockwork universe of early-nineteenth-century physics." (p. 77).

Kuhn, Thomas S. *The Structure of Scientific Revolutions.* Chicago: University of Chicago Press, 1962.

This book has become a classic. It has had a profound effect on work in the history and philosophy of science, and on many other areas in the social sciences. Kuhn traces the major upheavals in science to the overthrow of paradigms or conventional world-models. Such paradigms as the Newtonian universe and the caloric theory of heat continued to prevail in the face of contradictory evidence and ultimately died out only with the passing of the generation of their devotees. Because of the work of Kuhn and Holton, the subjective element in science can never again be ignored.

Needham, Joseph. *Science and Civilization in China.* 5 vols. Cambridge: Cambridge University Press, 1954-74. See also *The Shorter Science and Civilization in China: 1*, abridged by Colin A. Ronan. Cambridge: Cambridge University Press, 1980.

An authoritative, scholarly, yet eminently readable survey and analysis of Chinese philosophy, science, and technology. Volume 2 is especially useful in tracing the roots of Chinese science to early Oriental thought.

Needleman, Jacob. *A Sense of the Cosmos.* New York: Dutton, 1976.

A very thoughtful account of the deep common roots of science and religion. Needleman makes us see that the quest for meaning, purpose, and possibly even salvation is as much a part of science as it is of religion. Science has turned its back on these issues at its own, and humanity's, peril.

Ouspensky, P. D. *Tertium Organum.* New York: Knopf, 1959.

Ouspensky's considerable mystical propensities were neatly balanced by his mathematical insights, and nowhere better than in this book. Ouspensky argues that the level of conscious awareness in higher creatures corresponds to the degree of their spatiotemporal perceptions. In humans, higher awareness or enlightenment corresponds to perceiving the fourth dimension. Hierarchical structures in psychology, morality, logic, science, and other fields are also seen to parallel the levels of dimensional sensibility. I owe a great debt to Ouspensky and his creative examples of spatial intuition which have served to illustrate related ideas in my own book.

Pauwels, Louis, and Jacques Bergier. *The Morning of the Magicians.* New York: Avon, 1968.

A revealing exploration of many ideas, theories, and phenomena that are discredited and rejected by science, including astrology, alchemy, the myth of Atlantis, the use of magic in early civilizations, and methods of attaining states of higher consciousness and awareness. In an illuminating chapter called "The Magic Mind Rediscovered," a perceptive contrast is made between digital and analogue knowledge which suggestively hints at the limits to linear rational understanding.

Pirsig, Robert M. *Zen and the Art of Motorcycle Maintenance*. New York: William Morrow, 1974.

A superb autobiographical account—told as a kind of modern-day myth—of one man's life-and-death struggle with the philosophical issues on which his own sanity hinges. Pirsig's musings on rationality and intuition are clear, lively, and provocative. He weaves them poetically into a symbolic story which reflects the debate in his mind and adds meaning and beauty to this book. A rare and masterful blend of philosophy and literature.

Polanyi, Michael and **Harry Prosch.** *Meaning*. Chicago: University of Chicago Press, 1975.

In this final book of Polanyi's there is a grand synthesis of his case against the objectivity of science and his views on the arts and society. Polanyi relates the essential human tacit element in all knowledge to the creative imagination. Order and meaning are created and celebrated in science as well as in art. The essential environment for the flowering of human genius and significance is a free society which Polanyi champions and reveals as the deep-seated motive behind his philosophical struggle against chaos and meaninglessness.

Roszak, Theodore. *Where the Wasteland Ends*. Garden City, N.Y.: Doubleday, 1973.

All of Roszak's books, from *The Making of a Counterculture* on, constitute a brilliant attack on the narrow-mindedness and idolatry of science and, at the same time, a sober analysis and appreciation of the best aspects of the mystic-romantic countermovement in our society. No one can blame Roszak (or Lewis Mumford) if scientists do not take heed. This book is especially valuable in tracing the blindness inherent in the Newtonian outlook. William Blake has found an able champion and prophet in Theodore Roszak.

Turbayne, Colin. *The Myth of Metaphor*. Columbia, S.C.: University of South Carolina Press, 1970.

A fascinating vivisection of idolatry. Turbayne makes it painfully clear that different metaphors may serve the same function and that our models of reality are therefore relative. The detailed presentation of a linguistic metaphor for vision beautifully illustrates this important idea. There is also an extremely valuable analysis of how Newton and Descartes led the way in attributing logical mental structures to the physical processes of nature.

Watts, Alan. *The Book*. New York: Random House, 1966.

The books of Alan Watts, taken as a whole, form a major invaluable introduction to Eastern thought and religion. In this book, Watts interprets for Westerners the ancient Hindu philosophy of Vedanta. It is an exploration of the self with the aim of shattering the notion of an individual ego, separated from the rest of the cosmos. In his more recent autobiography, *In My Own Way*, we are given a charming view of Watts's lifelong quest for this elusive goal.

Whorf, Benjamin L. *Language, Thought, and Reality*. Cambridge, Mass.: M.I.T. Press, 1956.

An important glimpse into alternative views of reality through the language and thought patterns of American Indians. The chapters on the Hopi experience of space and time are especially valuable.

Zuckerkandl, Victor. *Sound and Symbol.*
Princeton, N.J.: Princeton University Press, 1969.

A profound "critique of our concept of reality from the point of view of music" (p. 364). Through its discussion of music as space, time, and motion, and its analysis of music as neither a purely inner nor outer experience, this book makes us confront a completely unfamiliar set of metaphors for reality and raises questions about meaning and symbol in science.

Index

Index

Roger S. Jones received his Ph.D. in physics at the University of Illinois in 1961 and then worked as a research scientist in the field of experimental high-energy physics at the Laboratorio Nazionale di Frascati in Italy and at the Brookhaven National Laboratory. In 1967, aware of his growing disenchantment with pure scientific research, Jones turned to teaching and assumed his current position as associate professor of physics at the University of Minnesota. Active in the formation of an Experimental College at Minnesota in the early 1970s, Jones developed courses with a humanistic approach to physics and mathematics and has also taught such courses in Minnesota's Humanities Program.